ACCADEMIA NAZIONALE DEI LINCEI

SCUOLA NORMALE SUPERIORE

LEZIONI FERMIANE

V.I. ARNOLD

The Theory of Singularities and Its Applications

PISA - 1991

Published for the Accademia Nazionale dei Lincei and the Scuola Normale Superiore by
the Press Syndicate of the University of Cambridge
The Pitt Building, Trumpington Street, Cambridge CB2 1RP
40 West 20th Street, New York, NY 10011, USA
10 Stamford Road, Oakleigh, Melbourne 3166, Australia

© Accademia Nazionale dei Lincei 1991

First published 1991

Printed in Italy by Pantograf

British Library cataloguing in publication data available

Library of Congress cataloguing in publication data available

ISBN 0 521 422809 paperback

CONTENTS

Preface — 3

1. – The Zoo of Singularities — 5

    1.1. Morse theory of functions — 5
    1.2. Whitney theory of mappings — 10
    1.3. The Whitney-Cayley umbrella — 14
    1.4. The swallowtail — 17
    1.5. The discriminants of the reflection groups — 21
    1.6. The icosahedron and the obstacle by-passing problem — 28
    1.7. The unfurled swallowtail — 32
    1.8. The folded and the open umbrellas — 36
    1.9. The singularities of projections and of the apparent contours — 39

2. – Singularities of Bifurcation Diagrams — 44

    2.1. Bifurcation diagrams of families of functions — 44
    2.2. Stability boundary — 52
    2.3. Ellipticity boundary and minima functions — 53
    2.4. Hyperbolicity boundary — 56
    2.5. Disconjugate equations, Tchebyshev system boundaries and Schubert singularities in complete flag manifolds — 57
    2.6. Fundamental system boundaries, projective curve flattenings and Schubert singularities in Grassmann manifolds — 60

References — 69

Index — 71

PREFACE

The mathematical description of the world depends on a delicate interplay between discrete and continuous objects. Discrete phenomena are perceived first, but continuous ones have a simpler description in the terms of the traditional calculus. Singularity theory describes the birth of discrete objects from smooth, continuous sources.

The main lesson of singularity theory is that, while the diversity of general possibilities is enormous, in most cases only some standard phenomena occur. It is possible and useful to study these standard phenomena once for all times and then to recognize them as the elements of more complicated phenomena, which are combinations of those standard elements.

In these lectures I shall describe some of such elementary singularities and some of the simplest of their applications. For instance, we shall consider the singularities of the domains in functional spaces, defined by such conditions, as the conditions of stability, of ellipticity, of hyperbolicity, of disconjugacy and so on, and the unexpected relations of singularity theory to reflection group theory (where, for instance, the variational problem of the by-passing of an obstacle, bounded by a smooth plane curve or space surface, is related to the icosahedron or to the hypericosahedron, that is to the regular-600-hedron in 4-space).

The results of singularity theory may be understood and used independently of their proofs, which are rather technical and involved. The proofs of the majority of the results, discussed below, may be found in the book "Singularities of Differentiable Mappings" by V.I. Arnold, S.M. Gusein-Zade, A.N. Varchenko, Moscow, Nauka, vol. 1, 1982, vol. 2, 1984 (English translation by Birkhäuser, vol. 1, 1985, vol. 2, 1988) and in the series Itogi Nauki i Techniki, Sovremennye Problemy Math., Novejshie Dostijenia, Moscow, Viniti, vol. 22 (1983) and vol. 33 (1988) (translated into English as J. Soviet Math. by Consultants Bureaux, Plenum, N.Y.; vol. 22 is translated in 1984 as part of the vol. 27 of the J. Sov. Math.). For more details on singularities, see the survey "Singularities" by V.I. Arnold, V.A. Vassiljev, V.V. Gorjunov and O.V. Ljashko in the series Itogi Nauki i Techniki, Sovremennye Problemy Math., Fundamentalnye napravlenia, Moscow, Viniti, vol. 6 (1988) and vol. 39 (1989) (translated by Springer as "Encyclopedia of Mathematical Sciences", Dynamical Systems volumes 6 and 8).

1. - THE ZOO OF SINGULARITIES

We shall start by showing a small set of singularities which occur in very different problems. These singularities are as fundamental as, say, the ellipse, the hyperbolas and the parabola. Their occurence in very different theories depends on the universality of the same kind as the universal occurence of quadratic forms in all branches of mathematics and physics.

1.1 - Morse theory of functions

Let us consider a function $y = f(x)$. (Here and always all the functions and mappings are supposed to be sufficiently smooth, say - infinitely differentiable or even analytical, that is developable into a Taylor series convergent to those functions or mappings at a neighbourhood of every point. The main part of the theory remains valid and nontrivial even in the case where all the functions are supposed to be polynomials.)

The singularities we shall discuss are *critical points*. For a typical function the critical points are non-degenerate. For functions depending on n variables (such that x belongs to the n-space $\mathbb{R}^n$ or to an n-dimensional manifold), a critical point of a function is called *nondegenerate*, if its second differential is a nondegenerate quadratic form.

MORSE LEMMA. *In a neighbourhood of a nondegenerate critical point, a function may be reduced to its quadratic part, i.e. it may be written in the normal form*

$$y = -x_1^2 - \ldots - x_k^2 + x_{k+1}^2 + \ldots + x_n^2$$

for a suitable choice of the local coordinate system $(x_1, \ldots, x_n)$, *whose origin is at the critical point (the origin at the axis of the values, y, is supposed to be choosen at the critical value).*

The Morse Lemma explains the occurrence of quadratic forms (and hence of ellipses, of hyperbolas and so on) in most of the problems of calculus, geometry and physics: they are the normal forms of *arbitrary* generic functions in the vicinity of their critical points. (Similar reasoning forms the *raison d'être* of the whole of algebraic geometry: polynomials are either local approximations or the local normal forms of arbitrary functions or mappings.)

In the functional space of all functions, degenerate functions form a hypersurface (a submanifold of codimension one, that is defined by one equation). This hypersurface is called the *bifurcation set*. The bifurcation set is not smooth: it consists generally of smooth manifolds of different dimensions, adjacent to each other in a very special way. The study of the structure of this bifurcation set (and of similar bifurcation sets in other problems) is the main content of singularity theory.

A typical function has only nondegenerate critical points, and its local structure is completely described by the Morse lemma, if we consider two functions equivalent when they are transformed one into another by a smooth change of the independent and of the dependent variables.

Globally, the set of critical points does not define a function up to equivalence (even in the case when we know the index k, which allows us to distinguish the maxima, the minima and different kinds of saddle-points). This is true even for functions of one variable.

PROBLEM. Find the number $K(n)$ of pairwise nonequivalent functions of one variable having n nondegenerate critical points with pairwise different critical values, supposing that at infinity the function behaves like x for even n and like x^2 for odd n.

ANSWER. $K = 1, 1, 1, 2, 5, 16, \ldots$ for $n = 0, 1, \ldots$;

$$\sum K(n) \frac{t^n}{n!} = \sec t + \tan t.$$

[REMARK (for the experts). This result shows, for instance, that Euler numbers and Bernoulli numbers together form a single sequence. It also opens the way for the definition of Euler and Bernoulli numbers associated to any simple Lie algebra (the usual case corresponding to unitary groups), or even to general singularities.]

The bifurcation set divides the function space into components. Two functions in the same component will be equivalent to each other if we include in the bifurcation set two parts: the hypersurface of functions having degenerate critical points and the hypersurface of functions with coinciding critical values. The first hypersurface is called the *caustic* (in the corresponding optical problem it is the place of the concentration of light). The second hypersurface is called the *Maxwell set* (because of the Maxwell rule in phase transition theory; this rule is the condition for the coincidence of two critical values of some function).

PROBLEM. Find the number of parts into which the full bifurcation set (consisting of functions with degenerate critical points or multiple critical values) divides the space of polynomials of the form

$$x^{n+1} + a_1 x^n + \ldots + a_{n+1}.$$

ANSWER. $K(n) + K(n-2) + K(n-4) + \ldots = 1, 1, 2, 3, 7, 19, \ldots$ for $n = 0, 1, \ldots$.

EXAMPLE. The plane of polynomials $x^5 - x^3 + ax^2 + bx$ is divided by the bifurcation set into 7 parts, shown in fig. 1.

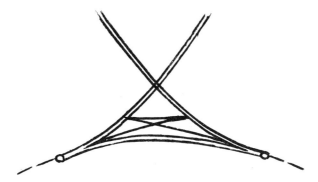

Fig. 1 - The caustic and the Maxwell stratum of a family of functions.

In this case the caustic (shown in fig. 1 as a double line) and the Maxwell set (completed by its analytical continuation, which is represented in fig. 1 by a hatched line) are diffeomorphic: these two plane curves can be transformed one into the other by a diffeomorphism of the plane (each curve has two cusps and one self-intersection point).

The decomposition of the bifurcation set into submanifolds of different dimensions is represented in fig. 1 by the decomposition of the curve into 6 points and 10 intervals.

A generic function has neither degenerate critical points nor multiple critical values. However, degenerate points and multiple critical values occur unavoidably in the families of functions depending on parameters.

Let us consider, for instance, one-parameter families. A one-parameter family is represented in function space by a curve. This curve may intersect the bifurcation hypersurface. If the intersection is transversal ("the angle between the curve and the hypersurface" being nonzero; and the intersection point being a generic point of the hypersurface), then the intersection is stable (it cannot be destroyed by a small variation of the family). Of course, for a neighbouring family, the intersection will happen for a slightly different value of the parameter, when compared with the non perturbed family, and the point of intersection itself is slightly different. But it is impossible to remove the intersection with the bifurcation set for all the values of the parameter simultaneously by a small variation of the family.

What can be achieved by a small variation of a one-parameter family is transversality: we may force the curve representing the family to intersect the bifurcation hypersurface in its generic points only and "at nonzero angles".

In fig. 1 a curve representing a typical one-parameter family, should not contain any of the 6 singular points of the bifurcation curve and should not be tangent (neither to the caustic curve nor to the Maxwell curve). Of course, those events (tangency or passage through a singular point) are possible for some special families. But we may destroy them by a small variation of the family.

The same way a k-parameter family is represented by a k-dimensional submanifold of the function space. A typical k-parameter family intersects only those parts (strata) of the natural decomposition of the bifurcation set whose codimension is k or smaller. And the intersections are transversal (no tangency). Hence the trace of the bifurcation set on the k-manifold, representing the family, provides a correct picture of the singularity of the bifurcation set at the stratum corresponding to the point of intersection. This trace is called the *bifurcation diagram of the family*. It may be considered as living in the parameter space of the family.

EXAMPLE. The two-parameter family of functions in fig. 1 is transversal to the bifurcation set in the function space. Fig. 1 provides a correct picture of the singularities of the bifurcation set at its strata of codimension 1 and 2.

Thus, in a typical one-parameter family of functions, degenerate critical points occur for some special values of the parameter. Those degenerate critical points correspond to a transversal intersection of the curve, representing the family in function space, with the caustic.

In a neighbourhood of such a point, the function of one variable may be reduced (by a smooth change of variables) to the normal form $y = x^3$.

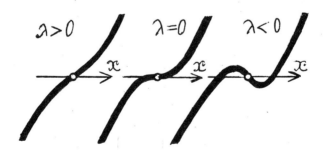

Fig. 2 - A perestroika of the birth of two critical points.

Moreover, the whole family may be reduced to the normal form of fig. 2,

$$y = x^2 + \lambda x$$

where λ is the parameter (the reduction is achieved by a smooth change of the parameter and a smooth change of coordinates, which depend smoothly on the parameter).

[In the case of more than 1 independent variable, to the normal form x^3 we have to add a nondegenerate quadratic form of the other variables, so the final normal form of the family is

$$y = -x_1^2 - \ldots - x_k^2 + x_{k+1}^2 + \ldots + x_{n-1}^2 + x_n^3 + \lambda x_n.]$$

Thus, the transversal intersection of a one-parameter family of functions with the caustic implies the birth or death of a pair of neighbouring critical points, whose indexes differ by one (whether the birth or the death takes place depends on the direction of the variation of the parameter).

In a typical two-parameter family, a further degeneration may occur. The corresponding surface may pass through the (simplest) singular points of the caustic. The corresponding normal form (for the families of functions of one variable) is

$$y = x^4 + \lambda_1 x^2 + \lambda_2 x$$

(if the number of independent variables is larger, one has to add a nondegenerate quadratic form of the other variables).

The bifurcation diagram of the family (the trace of the caustic on the plane of the parameters (λ_1, λ_2)) is a semicubical parabola $\Delta = 0$, where $\Delta = 8\lambda_1^3 + 27\lambda_2^2$. It means that the bifurcation set in function space has a codimension two stratum (the "cusped edge") of a semicubical type. (The adjacent bifurcation hypersurface intersects the two-dimensional surface transversal to the codimension two stratum along a curve having a semicubical cusp point - the bifurcation diagram, also called in this case the *caustic* of the family, which is represented by the surface.)

For typical 3-parameter families a further degeneration occurs stably (corresponding to the codimension 3 stratum of the caustic). The corresponding normal form of the family (of functions of one variable) is

$$y = x^5 + \lambda_1 x^3 + \lambda_2 x^2 + \lambda_3 x.$$

The bifurcation diagram (the caustic), in the space of parameters $\{(\lambda_1, \lambda_2, \lambda_3)\}$, is a surface with a singularity shown in fig. 3, studied by Kroneker and called the *swallowtail*.

Fig. 3 - The swallowtail.

Its section by the plane $\lambda_1 = -1$ is represented in fig. 1. It is interesting to note that the trace of the Maxwell set (continued analytically) is diffeomorphic to the trace of the caustic: the trace of the Maxwell set in the space of parameters $(\lambda_1, \lambda_2, \lambda_3)$ is also a swallowtail (compare fig. 1).

The topological properties of different strata of the natural stratification of the space of functions on a manifold are, in principle, the invariants of the smooth structure of the manifold. But very little is known about these topological properties.

EXAMPLE. The space of functions on a line, coinciding with x at infinity and having no critical points more complicated than x^3, is weakly homotopy equivalent to the loop space of the two-sphere, ΩS^2.

Let us associate to any point x the vector $(f'(x), f''(x), f'''(x))$ of 3-space. Since this vector is nowhere zero, we obtain a mapping from the axis of x to the two-sphere. The conditions at infinity imply that this mapping defines a loop. We have thus constructed a mapping from our function space to the space of loops. This mapping happens to be a weak homotopy equivalence.

REMARK (for the experts). Thus, in this case, the "principle of nonrelevance of integrability conditions" of Smale-Gromov holds, while this does not follow from any known general theorem (of course, replacing x^3 by x^n we have to substitute S^{n-1} to S^2, see [1]).

1.2 - Whitney theory of mappings

H. Whitney found in 1955 the local normal form of the singularities of typical mappings from two-dimensional manifolds to the plane (or to another two-dimensional manifold):

$$y_1 = x_1^3 + x_2 x_1, \quad y_2 = x_2.$$

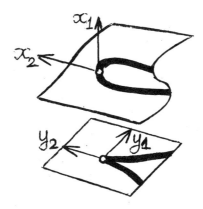

Fig. 4 - The Whitney tuck mapping.

EXAMPLE. Let us consider the surface, defined by the first equation (fig. 4) in the 3-space with coordinates (x_1, x_2, y_1). Its projection to the (y_1, x_2) plane along the x_1 axis has the required singularity. This singularity is stable (any neighbouring mapping has, at some neighbouring point, an equivalent singularity, i.e. a singularity which may be reduced to the same normal form by some smooth change of the independent and of the dependent variables).

The Whitney formula may be considered as defining a one-parameter family of functions $y_1(x_1)$, depending on $x_2(= y_2)$ as a parameter. This family experiences a perestroika of fig. 2 at the value 0 of the parameter (two critical points are born when the parameters move from the positive to the negative values).

The set of critical points of the Whitney mapping (where its jacobian vanishes) is a smooth curve (a parabola on the plane of independent variables x). The set of critical values is a curve with a semicubical cusp on the plane of the dependent variables y. For the projection mapping of fig. 4, the critical points are the points of the surface where the tangent plane is vertical, and the critical values form the "apparent contour" of the surface seen from above (we suppose the surface to be transparent).

The apparent contours of the generic smooth surfaces (projected along the typical directions) have no other singularities besides the Whitney cusps (and the obvious self-intersections). In particular, one can find the cusps at the apparent contours of the faces of people (fig. 5), but we usually do not perceive them since the faces are not transparent.

Fig. 5 - A singularity of an apparent contour.

A semicubical singularity is usually a sign of a Whitney mapping hidden somewhere in the neighbourhood. The generic singularities of typical mappings of three-dimensional manifolds to the plane may be considered as being the combinations of Morse and Whitney singularities. The normal form is

$$y_1 = x_1^3 + x_1 x_2 + x_3^2, \quad y_2 = x_2.$$

The set of critical values is still the semicubical parabola in the plane of dependent variables y. The preimages of the values of y, which do not lie on this parabola, are smooth elliptic curves on the planes $\{x_1, x_2, x_3 : x_2 = \text{const}\}$. The variables y may be considered as presenting the parameters of this family of curves. The semicubical parabola of the critical values is the bifurcation diagram of this family of curves.

In the same way, the semicubical parabola of the critical values of the Whitney mapping from a surface to the plane may be viewed as being the bifurcation diagram of the corresponding family of the 0-dimensional subsets of the line (the subsets being the intersections of the vertical lines, along which we project the surface, with the surface).

EXAMPLE. Let us consider the family of normals to an Euclidean plane curve, say, to an ellipse (fig. 6).

Fig. 6 - The caustic of an ellipse.

This family has an envelope (which is an astroide in the case of an ellipse). The envelope (called also the *caustic* of the curve, its *evolute*, its *focal set* and its *curvature centre set*) has, generically, singular points - semicubical cusps. These singularities of the caustics are stable: they do not disappear under a small variation of the initial generic smooth curve, they move smoothly with it.

As semicubical cusps are the signs of Whitney singularities, we seek for the corresponding surface mapping. In this case, it is the *normal mapping* associated to the curve. The source surface is the manifold of all the vectors normal to the curve. This 2-manifold is called the *normal bundle space* of the curve. The normal mapping associates to a normal vector (considered as being an arrow in the plane starting from a point of the curve) its end point.

A critical point of the normal mapping is a vector joining a point of the curve with the curvature centre of the curve at that point. The critical values form consequently the curvature centre set of the curve. But the critical values of the normal mapping are also the intersection points of infinitesimally close normals of the curve. Hence, the critical value set of a generic curve is the envelope of its normals.

The normal mapping of a generic curve has only the singularities typical for the generic mappings between two-dimensional manifolds, that is only the Whitney singularities: the "folds" have the normal form

$$y_1 = x_1^2, \ y_2 = x_2$$

displayed as the caustic line, and the "cusps" (or "tucks") of fig. 4, displayed as the cusps of the caustic line.

WARNING. The genericity of the singularities of the normal mapping associated to a generic curve is not evident. Indeed, the normal mappings associated to curves form a narrow subclass (of infinite codimension) in the functional space of all the surface mappings. This might lead, in principle, to the following two phenomena. 1°. Some singularities, typical for the general surface mappings, might become nontypical (or even impossible) for normal mappings. 2°. Some singularities, typical for normal mappings, might become nontypical for general surface mappings (they might be destroyed by arbitrary small perturbations in the class of general surface mappings, but not in the class of the normal mappings).

While neither of these two phenomena does occur for the normal mappings associated to plane curves, the situation in higher dimensions is different. Indeed, the typical singularities of the caustics (or of the curvature centre varieties) of dimension 2 or more, differ from the typical singularities of the critical sets of the generic mappings of the manifolds of the corresponding dimension (n for the curvature centre varieties of the hypersurfaces in Euclidean or Riemannian n-spaces). The singularities of the normal mappings are better understood in the theory of the Lagrange singularities of symplectic geometry [2], [3].

1.3 - The Whitney-Cayley umbrella

The normal form of a typical mapping of a surface $\{(u,v)\}$ into the three-space $\{(x,y,z)\}$ was also found by Whitney:

$$x = u, \quad y = uv, \quad z = v^2.$$

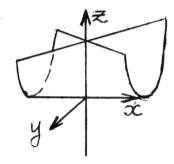

Fig. 7 - The Whitney-Cayley umbrella.

The image of this mapping is the surface $y^2 = zx^2$ (fig. 7), called the Whitney umbrella (or the Cayley umbrella, since this and most of the other important singularities first appeared in the works of Cayley).

Strictly speaking, this eccentric umbrella (which is not really any use in rain) also contains a handle $\{x = y = 0, z < 0\}$, which is not included in the image of the real plane $\{(u,v)\}$ under the above Whitney mapping.

This mapping has just one critical point (the origin). The critical value is also the origin ($x = y = z = 0$). The other points of the (positive) z axis are the points where a transversal intersection of two "smooth branches" of the umbrella surface occurs. Each self-intersection point has two preimages at the plane of the independent variables (u,v). In a neighbourhood of each of these preimage points, the mapping is nondegenerate (it defines a local embedding of a plane in 3-space).

It is interesting to note that the surfaces of the swallowtail (fig. 3) and of the umbrella (fig. 7) are homeomorphic (topologically equivalent). Moreover, there exists a homeomorphism (a one-to-one continuous mapping whose inverse mapping is also continuous) of the ambient 3-space, transforming the swallowtail surface into the umbrella surface.

This homeomorphism is a manifestation of general multidimensional phenomena (B.A. Hessin, B.Z. Shapiro, 1989).

The umbrella surface is related to the perestroikas of the projections of the space curves to the plane. A typical space curve (the image of a generic mapping of a line into 3-space) has no singularities (it is a smooth embedded submanifold). A generic projection of such a curve to the two-dimensional plane may have some points of transversal self-intersection (where two branches meet at a nonzero angle). One can get rid of all the more complicated singularities of the projection by a small variation of the direction along which we project our curve.

Suppose now that the direction of the projection depends on a parameter. Then, for some values of the parameter, a more complicated singularity of the projection may occur: a semicubical cusp point on the projection curve. Indeed, the manifold of all directions of the projection in 3-space is two-dimensional. The set of tangential directions of a generic space curve is one-dimensional (it forms a curve on the projective plane of the directions). Our one-parameter family of the directions of the projections is another curve on the same projective plane. A typical family is a generic curve, and it may intersect transversally the curve formed by the tangents of the space curve. The intersection point corresponds to the projection along a tangent of the space curve. The curvature and the torsion of the space curve at the point of tangency do not vanish (for a generic space curve and a generic family of projection directions). Hence the projected curve has a semicubical cusp point.

Let us consider now the projections which correspond to the neighbouring directions of the projections contained in our one-parameter family (fig. 8).

Fig. 8 - A perestroika of the projections of a plane curve.

Varying the parameter to one side of the bifurcation value (corresponding to a projection along a tangent line of the space curve), we see that the singularities of the projection disappear completely; varying to the other side, the cusp point is replaced by a selfintersecting loop.

Now let us consider the collection of these projected plane curves, corresponding to different values of the parameter, as belonging to different horizontal planes, numbered by the values of the parameter (equal to the height of the plane). These curves in different horizontal planes form together a surface in 3-space. This surface has a singularity which is a Whitney umbrella. The corresponding mapping, from a plane to the 3-space, sends the pair formed by a value of the parameter and a point of the space curve, to the projection of this point (along the direction corresponding to the value of the parameter) to the horizontal plane situated at the height equal to the value of the parameter.

It is possible to reverse these arguments. Starting from fig. 7, we consider a generic function on 3-space, where the standard umbrella $y^2 = zx^2$ surface lives, (for instance, the function $f = x+y+z$). The sections of the umbrella surface, by the level surfaces of the function, are curves on the 2-surfaces, depending on the value of the function as well as on the parameter. When the parameter varies, these curves experience (for the parameter value corresponding to the vertex of the umbrella) the same "$u \to \gamma$" perestroika, as do the typical projections of a generic space curve (fig. 8). One can prove that a generic function may be locally reduced to the normal form $x+y+z$ + const by an umbrella-preserving diffeomorphism of 3-space (in a neighbourhood of the vertex of the umbrella).

The vertices of the umbrella surfaces are also called the "pinch points" in physics and simply "vertices" in algebraic geometry.

PROBLEM. Let us consider a typical family of "plane curves", depending on a parameter and subjected to the "$u \to \gamma$" perestroika of fig. 8 when the parameter varies. Let us associate to each curve the union of its tangents at the inflection points (considering each curve, as above, as living in its own horizontal plane, at the height equal to the value of the parameter).

Prove that the surface formed by these tangents (and by their limiting positions) in 3-space, has a singularity, diffeomorphic to the standard umbrella (i.e. it can be transformed to the standard normal form by a local smooth change of variables in 3-space).

PROBLEM (see [4]). Let us consider the space of homogeneous polynomial plane vector fields, $P\frac{\partial}{\partial x} + Q\frac{\partial}{\partial y}$, where P and Q (the components of the field) are two homogeneous polynomials of degree $n \geq 2$ in x and y. Such a field is *degenerate* if it has a singular point other than the origin.

The set of the degenerate fields form a *bifurcation hypersurface* in the space of fields. This hypersurface has singularities (at those points where the forms P and Q have more than one common zero on the projective line, the polynomials P and Q vanishing together everywhere more than one line, counting the zeroes with their multiplicities).

Let us consider the singularity corresponding to a common root of multiplicity 2. Prove that the corresponding stratum of the bifurcation hypersurface has codimension 3 in the space of fields and that the intersection of the bifurcation hypersurface with a 3-dimensional space, transversal to this stratum, is a surface in 3-space having a singularity diffeomorphic to the standard umbrella at the point of the stratum.

HINT. Consider the case of homogeneous polynomials of degree $n = 2$.

REMARK. A similar construction defines the "higher umbrellas" (they correspond to the common roots of higher multiplicities). From the algebraic point of view, the (higher) swallowtail singularities correspond to the zeroes of the discriminants, while the (higher) umbrella singularities correspond to the zeroes of the resultants.

The occurrence of umbrellas in different problems of bifurcation theory is sometimes rather unexpected. Thus, umbrellas were found in a series of numerical experiments during the study of bifurcations of the 2π-periodical solutions of the equation

$$u_{xxxx} + A u_{xx} + u + u_x^2 = 0, \ A \approx 2.$$

The recognition of the umbrella has led to an explanation of the peculiar bifurcations observed in the experiments and to a general theory of bifurcations of the Ljapounov surfaces at the resonances in reversible dynamical systems (for more details see [5] and Sevrjuk's book [6]).

The strategy of using singularity theory in applications follows, in most cases, the same pattern: the geometry of the singularity (in particular the shape of the bifurcation diagram) is used as some kind of "fingerprint" of the singularity. After the recognition of the singularity by such fingerprints, the general theory usually suggests the simplest model in which such a singularity or bifurcation diagram occurs. The understanding of the observed phenomena attained this way, is frequently sufficient for the development of a quantitative theory, in which the theorems of mathematical singularity theory provide some explicit information on the phenomena under study (or at last on their models).

1.4 - The swallowtail

This surface (fig. 3) - one of the more common objects in singularity theory - has several dozen equivalent definitions. Continuing the description of the Whitney theory of singularities of mappings between manifolds of different dimension, we have to consider generic mappings between 3-manifolds.

The simplest singularity here is the "fold", which has the normal form

$$y_1 = x_1^2, \quad y_2 = x_2, \quad y_3 = x_3.$$

At the points of some curves, a more complicated "tuck" singularity occurs (three pre-images points collapsing together), having the normal form

$$y_1 = x_1^3 + x_1 x_2, \quad y_2 = x_2, \quad y_3 = x_3.$$

Finally, at some particular points, the most complicated singularity occurs, having the normal form

$$y_1 = x_1^4 + x_1^2 x_2 + x_1 x_3, \quad y_2 = x_2, \quad y_3 = x_3.$$

The set of the critical points of this mapping from a 3-space to a 3-space is a smooth surface (its equation being $4x_1^3 + 2x_1x_2 + x_3 = 0$).

PROBLEM. Prove that the set of the critical values of this mapping is diffeomorphic to the swallowtail surface.

PROBLEM. Prove that the following surfaces are all diffeomorphic to the swallowtail surface (i.e. can be transformed each into the other by a smooth change of variables in the ambient 3-space):

1. The set of polynomials $x^4 + ax^2 + bx + c$, having a multiple root.

2. The union of the tangents to the space curve $A = t^2$, $B = t^3$, $C = t^4$.

3. The set of polynomials $x^5 + ax^3 + bx^2 + cx$, having coinciding critical values.

(Strictly speaking, we should either consider the analytical continuation in the last case, or we should consider complex surfaces in all the 3 cases.)

Here are some other constructions which provide the swallowtails.

Let us consider a wavefront of some perturbation on a plane propagating with unit velocity from some initial position (for instance, a front of some perturbation, propagating inside an ellipse from its boundary).

The position of the front at time t is the equidistant curve of the critical curve. To obtain the equidistant curve, we consider the (oriented) normals of the initial curve. The equidistant curve is formed by the points at a distant t in the forward direction from the points of the initial curve along each normal. For small t the equidistant curve is smooth. But from some value of t on (namely, this critical value being the minimal curvature radius of the curve), the equidistant curve acquires singularities. For a typical initial curve, two semicubical cusp points are born on the equidistant curve (fig. 9).

Fig. 9 - *A perestroika of the equidistant curves of a plane curve.*

Let us consider now the whole family of the equidistant curves, lifting each curve at the height t in its own horizontal plane. We obtain a surface which is the graph of the "multivalued time function" $t(x, y)$ - the distance to the initial curve along any of its normals form the given point of the plane (this function is multivalued because there may exist several normals).

The graph of the multivalued distance function to a generic plane curve is diffeomorphic to the swallowtail surface in a neighbourhood of the curvature centre which is the closest to the curve (for instance, in a neighbourhood of the focus of an ellipse).

Reversing the reasoning, we can start with the swallowtail surface

$$\{x^4 + ax^2 + bx + c \text{ having a multiple root}\}$$

and can consider a generic function $f(a, b, c)$. The level surfaces f = const of this function intersect the swallowtail surface along the lines.

A generic function is reducible to the normal form $\pm a$ + const in a neighbourhood of the swallowtail vertex by a swallowtail preserving diffeomorphism of the ambient space. Hence the decomposition of the swallowtail surface into the level lines of a generic function is diffeomorphic to its decomposition in the intersections with the vertical planes a = const (fig. 10).

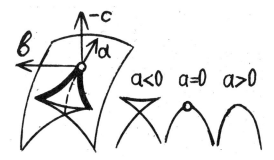

Fig. 10 - *A perestroika of the plane sections of a swallowtail surface.*

The decomposition of the graph of the multivalued time function into the level sets of the time (i.e. into the equidistant curves of the initial generic curve) is thus locally diffeomorphic to the decomposition of the swallowtail, shown in fig. 10.

The equidistant surfaces of a surface in Euclidean 3-space also have singularities. For a generic surface (for instance, for a tri-axial ellipsoid) these singularities (studied by Cayley) are the semicubical cusped edges and the swallowtails (we do not distinguish the locally diffeomorphic surfaces and we do not take into account self-intersections).

It should be noted that an equidistant surface of a smooth surface is the image of a smooth mapping from the last surface to Euclidean space. Namely, this equidistant mapping sends a point of the surface into the point of the normal to the surface (at the initial point) at the distance t from this initial point. However, the singularities of the equidistant surfaces of the generic surfaces (the cusped edges and the swallowtails) are very different from the singularities of the generic mappings from surfaces to 3-space (the umbrellas).

The singular points of the typical mappings are some isolated points on the source surfaces, while the singular points of the equidistant mappings form the curves on the source surfaces (these curves are the level lines of the principal curvatures). This is not astonishing, since the equidistant mappings form a very narrow subclass (of infinite codimension) in the functional space of all mappings. The equidistant mapping singularities are included in a special theory of Legendre singularities in contact geometry (which describes also the typical singularities of the Legendre transformations, of wave fronts, of the dual hypersurfaces, of the smooth projective hypersurfaces, and so on).

The caustics (the curvature centre sets) of the typical surfaces of Euclidean 3-space also have at some points singularities diffeomorphic to the swallowtail. But in this case this is not the only typical point singularity (as it is the case for the wave fronts).

A typical caustic in 3-space has (besides the semicubical cusped edges) point singularities of three types: swallowtails, pyramids and purses (fig. 11).

Fig. 11 - The pyramid and the purse - typical singularities of caustics.

At the vertex of the pyramid, three (quadratically tangent to each other) cusped edges meet. At the purse singularity vertex, two diffeomorphic parts of the caustic intersect each other. Each of these two parts has a cusped edge forming a smooth ray issuing from the purse vertex (these two rays form together a smooth curve).

The purse singularity is the singularity of the caustic of a tri-axial ellipsoid at its umbilical point (where the two principal curvatures of the ellipsoid surface coincide).

In complex space the pyramid and the purse are diffeomorphic: these surfaces are two different real forms of the same complex surface (like the ellipse and the hyperbola are two different real forms of the same complex curve). In particular, the purse surface has three cusped edges, but two of them do not emerge in real space.

Swallowtails, pyramids and purses are the stable singularities of the caustics. The focal set of a sphere consists of its centre, which is a surface singularity of a different kind. But this singularity of the caustic is unstable. A small generic deformation of the sphere (for instance, its deformation to a triaxial ellipsoid close to a sphere) transforms the focal set into a small surface, whose singularities are cusped edges, swallowtails and purses (fig. 12).

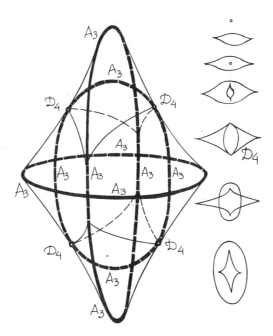

Fig. 12 - The caustic of an ellipsoid.

1.5 - The discriminants of the reflection groups

The discriminants of the Euclidean reflection groups are one more source of the singularities.

A *reflection* in a linear Euclidean space is an orthogonal transformation, fixing all the points of a hyperplane (which is called the mirror) and reversing the normal line of this hyperplane.

Suppose we have several hyperplanes in Euclidean space. The reflections defined by these mirrors generate a group which may be finite or infinite.

EXAMPLE. Two mirrors in the plane generate a finite reflection group if the angle between the mirrors is commensurable with 2π, and an infinite group if it is incommensurable with 2π.

Any finite group generated by Euclidean reflections in the plane (in short, a reflection group) is the symmetry group of a regular n-gon. It is denoted by $I_2(n)$ (where 2 is the dimension of the plane).

All the finite reflection groups were classified by Coxeter. Before we formulate his results, let us consider the main

EXAMPLE (*the series A_k*). Let us start with the group of permutations of the coordinates, acting in Euclidean n-space. This reflection group is generated by reflections in the mirrors $x_i = x_j$ (it is even sufficient to take only the $n-1$ mirrors, for which $i - j = 1$). But this action is reducible: it leaves invariant the diagonal line $x_1 = \ldots = x_n$ and hence its orthocomplement hyperplane $x_1 + \ldots + x_n = 0$. Thus we obtain a Euclidean reflection group acting in this hyperplane $\mathbb{R}^k (k = n-1)$. This group is irreducible (there exists no subspace invariant under all the transformations of the group). This reflection group is denoted by A_k. Its mirrors are the intersections of the diagonal hyperplanes $x_i = x_j$ with the hyperplane $x_1 + \ldots + x_n = 0$.

For instance, A_2 is the group of symmetries of a regular triangle in the plane, $A_2 \sim I_2(3)$ (fig. 13), A_3 is the group of symmetries of a regular tetrahedron in Euclidean 3-space, and so on.

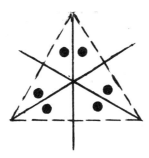

Fig. 13 - *The mirrors and an orbit of the reflection group A_2.*

Let us consider the action of a reflection group on a point. The orbit of the point may consist of at most as many points as the number of the elements in the group. Such a large orbit is called *regular* (and is shown in fig. 13). An irregular orbit consists of a smaller number of points. The orbits of the points belonging to the mirrors are irregular (see fig. 13).

Now let us consider the space of all orbits of a finite reflection group. A function (say, a polynomial) defined on our Euclidean space is called an *invariant* if it is invariant under the transformations forming the group. The values of an invariant, at all the points of the same orbit, coincide. Hence the invariants may be viewed as being functions on the space of orbits.

EXAMPLE. The invariants of the group A_2, acting on $\mathbb{R}^2$ as it is described above, are the symmetric functions of x_1, x_2, x_3, restricted to the plane $x_1 + x_2 + x_3 = 0$. All the invariant polynomials (restricted to this plane) may be represented as the polynomials of the $k(=2)$ *basic invariants* ($\sigma_2 = x_1x_2 + x_2x_3 + x_3x_1$, $\sigma_3 = x_1x_2x_3$), and this representation is unique. The invariants are constant on the orbits. The values of the basic invariants σ_2 and σ_3 define unambiguously the orbit (which is the set of roots of the polynomial $x^3 + \lambda_1 x + \lambda_2$, $\lambda_1 = \sigma_2$, $\lambda_2 = -\sigma_3$).

Now let us consider the complex values of all these variables. Then each (ordered) set of k values of the coefficients λ of the polynomial $x^{k+1} + \lambda_1 x^{k-1} + \ldots + \lambda_k$ defines the (unordered) set of its roots, that is an orbit of the action of the group A_k on the space

$$\mathbb{C}^k = \{x \in \mathbb{C}^n : \sum x_j = 0\}.$$

The correspondence between the sets of roots and (ordered) sets of coefficients is one-to-one. It is continuous in both directions. Thus we may identify the space of (complex) orbits of the group A_k with the space of polynomials

$$\mathbb{C}^k = \{x^{k+1} + \lambda_1 x^{k-1} + \ldots + \lambda_k\}.$$

Thus we have equipped the space of orbits with a structure of a smooth complex manifold, $\mathbb{C}^k$.

The basic invariants define now a polynomial *Vieta mapping*

$$\sigma : \mathbb{C}^k \to \mathbb{C}^k, \quad \lambda_1(x) = x_1x_2 + x_1x_3 + \ldots + x_{n-1}x_n, \quad \lambda_k(x)$$
$$= \pm\, x_1 \ldots x_n,$$

which sends the space where the reflection group A_k acts to the space of its orbits.

EXAMPLE. For $k = 2$, we obtain a mapping from the plane of the ordered triples of roots of the polynomials $x^3 + \lambda_1 x + \lambda_2$ to the plane of the coefficients (fig. 14).

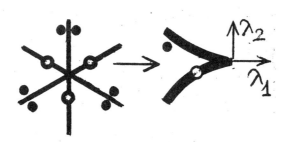

Fig. 14 - The Vieta mapping and the discriminant of the reflection group A_2.

The critical set of the Vieta mapping is the union of the mirrors. The critical values form the variety of polynomials having a multiple root. This variety is the (complexified) variety of the irregular orbits of the reflection group A_k.

DEFINITION. The variety of the irregular orbits of a reflection group is called its *discriminant*.

EXAMPLE. The discriminant of the group A_2 is the semicubical parabola in the plane (fig. 14), of the group A_3 - the swallowtail surface (fig. 3).

PROBLEM. Which of the 3 parts, into which the swallowtail surface cuts the real 3-space, is the image of the real 3-space under the Vieta mapping?

ANSWER. The smallest part, that is the interior of the pyramid of the swallowtail of fig. 3.

The discriminant of the reflection group A_k is the hypersurface, defined by the equation $\Delta(\lambda) = 0$, where Δ is the discriminant of the polynomial $x^{k+1} + \lambda_1 x^{k-1} + \ldots + \lambda_k$.

For any Euclidean reflection group in $\mathbb{C}^k$, the ring of invariant polynomials is a ring of polynomials of k basic invariants (this is a generalization of the basic theorem on symmetrical functions). Hence the orbit space is isomorphic to $\mathbb{C}^k$.

The variety of irregular orbits is a hypersurface. The Vieta mapping $\sigma : \mathbb{C}^k \to \mathbb{C}^k$ (associating to a point its orbit) sends the union of the mirrors onto the variety of irregular orbits. This hypersurface is also the critical value set of the Vieta mapping.

EXAMPLE (*series B_k*). Let us consider, in the space with coordinates $(x_1, \ldots, x_k)$, the group generated by the reflections in the coordinate hyperplanes $x_j = 0$ and in the diagonal hyperplanes $x_i = x_j$.

The orbit space is the set of polynomials $x^k + \lambda_1 x^{k-1} + \ldots + \lambda_k$. The Vieta mapping is defined by the formula

$$\lambda_1 = -(x_1 + \ldots + x_k), \ldots, \lambda_k = \pm \, x_1 \ldots x_k.$$

The discriminants of the groups B_2 and B_3 are shown in fig. 15.

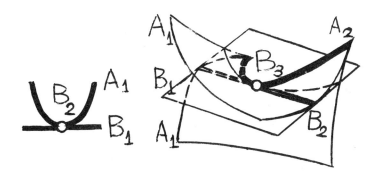

Fig. 15 - The discriminants of the reflection groups B_2 and B_3.

These hypersurfaces are reducible (each of them consists of two irreducible components). Indeed, there are mirrors and irregular orbits of two kinds: those consisting of polynomials with a zero root (B_1, $x_j = 0$) and those with a multiple root (A_1, $x_i = x_j$).

The component B_1, formed by the polynomials having a zero root, is smooth (since it is the plane $\lambda_k = 0$). The component A_1, formed by the polynomials having a multiple root, is singular when $k > 2$: it is diffeomorphic to a cylinder over the discriminant of the group A_{k-1}.

For instance, the discriminant of the group B_3, shown in fig. 15, contains a component, diffeomorphic to a cylinder over a semicubical parabola (the translations along the line, generating the cylinder, correspond to the shifts of the origin of the x axis). The smooth component of this discriminant is the tangent plane of the singular component at the origin.

The singularities shown in fig. 15 are fundamental for many applications of singularity theory.

PROBLEM. Draw the graphs of the polynomials corresponding to the 6 domains into which the 3-space of orbits of the group B_3 is cut by the discriminant. Find the image of the Vieta mapping acting on the real space $\mathbb{R}^3$.

ANSWER. The smaller part of the space bounded by the singular component and containing 4 of the domains.

The finite irreducible Euclidean reflection groups form two sets: the *crystallographic groups* (which preserve some lattices, i.e. some discrete subgroups of rank k in k-dimensional Euclidean spaces) and the *noncrystallographic groups*.

EXAMPLE. The groups A_k and B_k are crystallographic (they preserve the lattice of points having integer coordinates). The groups $I_2(n)$ of symmetry of regular n-gons are crystallographic for $n = 3, 4, 6$ and only for these values of $n > 2$. For instance, the pentagonal symmetry group is noncrystallographic, since no plane lattice admits symmetries of order 5.

To describe the reflection group, it is sufficient to describe the generating mirrors. For this it suffices to prescribe the angles between the lines normal to the mirrors. The following notation (called the Dynkin diagram, although it was used earlier by Coxeter and even earlier by Witt) is used: the basic vectors are denoted by the vertices of a graph, the angles by the edges. Explicitly an angle of 120° between two vectors is represented by an edge, joining the corresponding two vertices, an angle of 135° is represented by a doubled edge, an angle of 150° by a triple edge; the absence of an edge means an angle of 90°.

EXAMPLE. The diagram o——o denotes two mirrors in the plane, forming an angle of 120°, i.e. generating the reflection group A_2.

Now we are ready to list the finite irreducible Euclidean reflection groups, describing them by their diagrams.

1) *The crystallographic groups*

Groups with simple edges (angles of 90° and 120°). They form 2 infinite series and 3 exceptional groups:

$$A_k \quad \circ\!\!-\!\!\circ\!\!-\!\!\circ\!\!-\cdots-\!\!\circ\!\!-\!\!\circ$$

$$D_k \quad \circ\!\!-\!\!\overset{\displaystyle\circ}{\underset{|}{\circ}}\!\!-\!\!\circ\!\!-\cdots-\!\!\circ$$

$$E_6 \quad \circ\!\!-\!\!\circ\!\!-\!\!\underset{|}{\overset{\displaystyle}{\circ}}\!\!-\!\!\circ\!\!-\!\!\circ$$
$$\circ$$

$$E_7 \quad \circ\!\!-\!\!\circ\!\!-\!\!\underset{|}{\circ}\!\!-\!\!\circ\!\!-\!\!\circ\!\!-\!\!\circ$$
$$\circ$$

$$E_8 \quad \circ\!\!-\!\!\circ\!\!-\!\!\underset{|}{\circ}\!\!-\!\!\circ\!\!-\!\!\circ\!\!-\!\!\circ\!\!-\!\!\circ$$
$$\circ$$

Groups with nonsimple edges (angles of 90°, 120°, 135°, 150°). They form one infinite series and two exceptional groups:

$$B_k \approx C_k \quad \circ\!\!=\!\!\!=\!\!\circ\!\!-\!\!\circ\!\!-\!\!\circ \ldots \circ\!\!-\!\!\circ$$

$$F_4 \quad \circ\!\!-\!\!\circ\!\!=\!\!\!=\!\!\circ\!\!-\!\!\circ$$

$$G_2 \quad \circ\!\!=\!\!\!\equiv\!\!\circ$$

There exists no other irreducible finite crystallographic Euclidean reflection groups.

REMARK. This list coincides with the list of the Weyl groups of the simple complex Lie groups ($A_k \sim SU(k+1)$, $B_k \sim O(2k+1)$, $C_k \sim Sp(2k)$, $D_k \sim O(2k)$). It coincides also with the classification of the simple complex Lie algebras, if we take into account that the same reflection group $B_k = C_k$ has two different structures of crystallographic reflection groups, i.e. preserves two nonequivalent lattices.

2) *The non-crystallographic groups*

There exists one infinite series
$I_2(n)$ = the group of symmetries of a plane regular n-gon
and two exceptional groups

$$H_3 \;\; \circ\!\!\overset{5}{-}\!\!\circ\!\!-\!\!\!-\!\!\circ \quad \text{the group of symmetries of an icosahedron}$$

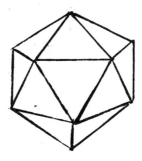

Fig. 16 - The icosahedron.

H_4 o—5—o—-o—-o the group of symmetries of a hyper-icosahedron.

The hyper-icosahedron is the regular 600-hedron in Euclidean 4-space. To construct it we start with the group of rotations of an icosahedron. This group consists of 60 elements of the rotation group $SO(3)$. Let us consider the two-fold "spin" covering $SU(2) \to SO(3)$. The preimage of the group of the icosahedron consists of 120 elements and it is called the *binary group* of the icosahedron.

Now $SU(2)$ is a 3-sphere. Hence the icosahedron binary group is formed by 120 points on the unit sphere of Euclidean 4-space. These 120 points are the vertices of a regular polyhedron, which is their convex hull and which we will call the hyper-icosahedron.

The group of its symmetries, H_4, is generated by reflections and consists of 120^2 elements (it is the product of the groups of the left and the right translations of the binary group). The number 5 on the H_k diagram means an angle of 144° (the pentagon symmetry group $H_2 = I_2(5)$ is defined by the diagram o—5—o).

1.6 - The icosahedron and the obstacle by-passing problem

The discriminant of the group H_2 is a plane curve with a singularity of order 5/2 (defined by the equation $x^2 = y^5$). This singularity is a cusp point, but it is not semicubical. It is easy to recognize this singularity in experimental data, since after a generic diffeomorphism such a curve consists of two branches that have equal curvatures at the common point and hence are convex from the same side (while, for a semicubical cusp, the two branches always have opposite convexities, fig. 17).

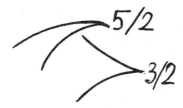

Fig. 17 - *The typical curves with cusps of order* 5/2 *and* 3/2.

The discriminant of the group H_3 is the surface shown in fig. 18.

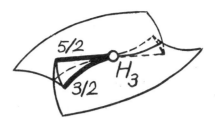

Fig. 18 - The discriminant of the symmetry group of an icosahedron.

Its singularities were studied by O.V. Ljashko (1982) with the help of a computer. This surface has two smooth cusped edges, one of order 3/2 and the other of order 5/2. Both are cubically tangent at the origin. Ljashko has also proved that this surface is diffeomorphic to the set of the polynomials $x^5 + ax^4 + bx^2 + c$ having a multiple root.

The comparison of this discriminant with the patterns of the propagation of the perturbations on a manifold with boundary (studied as early as in the textbook of L'Hôpital in the form of the theory of evolutes of plane curves), has led A.B. Givental to the conjecture (later proven by O.P. Shcherbak) that this discriminant is locally diffeomorphic to the graph of the multivalued time function in the plane problem on the shortest path, on a manifold with boundary, which is a generic plane curve.

Thus, the propagation of the waves, on a 2-manifold with boundary, is controlled by an icosahedron hidden at the inflection point of the boundary. This icosahedron is hidden, and it is difficult to find it even if its existence is known.

The problem of the fastest by-passing of an obstacle is the problem of the singularities of the (multivalued) distance function to a fixed point (or to a fixed submanifold) of a manifold with boundary.

EXAMPLE. Let us consider an obstacle bounded by a convex curve on the Euclidean plane. The level lines of the distance function are Huygens involutes of the boundary curve (as it follows from the very definition of the involutes, using the Huygens thread construction, shown in fig. 19).

Fig. 19 - An involute of a plane curve.

REMARK. Thus, we obtain a multidimensional generalization of the theory of the involutes: the involutes of a hypersurface are the fronts of the waves, propagating in the domain, bounded by the hypersurface.

Of course, when studying the involutes (like in the general theory of wave front singularities) analytical continuation is very useful. The continued involutes in fig. 19 contains also the hatched "nonphysical" part (corresponding to a negative length of the free part of the thread). We see now, that the involutes have a singularity at the boundary of the obstacle (namely, a semicubical cusp point, if the curvature of the obstacle curve does not vanish). This semicubical singularity was discovered by Huygens (who used it in his construction of an isochronous pendulum).

Let us consider now the evolutes of a curve that has the simplest inflection point (for example, the involutes of cubical parabola).

PROBLEM. Prove that the generic involute of a cubical parabola has a cusp of order $5/2$ on the straight line tangent to the parabola in the inflection point (fig. 20).

Fig. 20 - A perestroika of the involutes of a curve at the curve inflection point.

HINT. The curvature centres of both branches of the involute, which meet at the point of the inflectional tangent, lie at the inflection point, hence both branches have the same convexity (they are both concave from the side of the inflection point of the boundary).

Comparing fig. 20 with fig. 18, it is easy to guess that the graph of the distance (or time) function, whose level sets are the involutes, is diffeomorphic to the discriminant of the icosahedron symmetry group H_3 (the proof of this fact is not easy at all).

PROBLEM. Let us consider the variety of line elements at those points of the plane which are not generic with respect to a curve (all the elements based at the points of the curve are nongeneric, as well as all the elements of the tangent lines based at any point of the plane). Prove that this variety is locally diffeomorphic to the discriminant of the group B_3 (fig. 15) in a neighbourhood of the element of the inflectional tangent of a generic plane curve, based at the point of inflection.

HINT. The natural projection, from the 3-dimensional manifold of all line elements to the plane, defines a mapping of the variety of nongeneric elements to the plane, projecting the whole smooth component onto the boundary curve and the cusped edge of the nonsmooth component to the inflectional tangent.

In the obstacle problem in a 3-space, the discriminant H_3 occurs as the typical singularity of the fronts at those points on the surface of the obstacle where the extremal direction is tangent to the asymptotical direction of the surface.

The discriminant of the group H_4 of the symmetries of the hyper-icosahedron is one of the singularities of the graph of the multivalued time function in the 3-dimensional obstacle problem. This singularity occurs at a point of a ray which is asymptotical for the obstacle surface at a parabolical point of this surface, where the direction of the extremal of the obstacle problem is asymptotical for the surface.

This H_4 discriminant is a 3-dimensional variety in 4-space. Hence it may be described as a perestroika of 2-dimensional surfaces in 3-space, depending on a parameter.

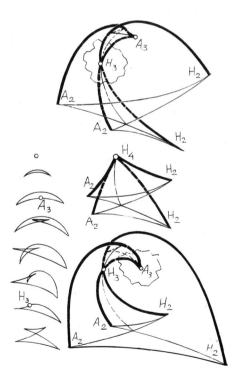

Fig. 21 - *The discriminant of the reflection group H_4.*

This perestroika is shown in fig. 21, borrowed, as the whole theory of H_4 singularities, from the paper of O.P. Shcherbak [7].

1.7 - The unfurled swallowtail

Let us consider, in the 4-dimensional space of polynomials

(*) $$x^5 + \lambda_1 x^3 + \lambda_2 x^2 + \lambda_3 x + \lambda_4,$$

the set of polynomials having a root of multiplicity at least three. This 2-dimensional surface is called the *unfurled* (or *open*) *swallowtail*.

The origin of this name is the following observation. Differentiation of the polynomials sends the 4-space of polynomials of degree 5 to a 3-space of polynomials of degree 4 (this mapping is a fibration with 1-dimensional fibres). The multiplicities of the roots decrease under the differentiation by one. Hence the surface of polynomials having a triple root is projected onto the swallowtail formed by the polynomials of degree 4 having a double root.

Thus differentiation defines a mapping of our surface onto the swallowtail surface. This mapping is a local diffeomorphism almost everywhere, the only exception being the self-intersection line of the usual swallowtail. The generic points of this line have two pre-images each.

Hence our surface in the 4-space has a cusped edge, like the usual swallowtail, but has no self-intersections. The lifting from the ordinary swallowtail surface to our surface in the 4-space eliminates the self-intersections of the ordinary swallowtail. That is the reason why this surface is called the open (or unfurled) swallowtail (fig. 22).

Fig. 22 - The projection of the open swallowtail onto the usual one.

PROBLEM. Prove that the variety of the (complex) polynomials (*), having at least two double roots, is diffeomorphic to the (complex) unfurled swallowtail.

PROBLEM. Prove that the real unfurled swallowtail is a continuous section of a smooth (polynomial) fibration of the 4-space over a 2-space (and, particularly, that the complement to the unfurled swallowtail is diffeomorphic to the complement of a 2-plane in $\mathbb{R}^4$).

The extremal curves in the problem of the fastest by-passing of an obstacle in Euclidean 3-space, are formed by the intervals of the geodesic lines on the obstacle surface and by the finite or infinite segments of straight lines tangent to those geodesics.

Let us consider the system of extremals defining the shortest (or at least the stationary) paths joining all the points of the space with a fixed initial submanifold (say, with a point). The geodesic lines included in the system form a one-parameter family of lines on the surface of the obstacle. The straight lines tangent to these geodesics, form a 2-dimensional variety in the 4-dimensional manifold of all straight lines of the 3-space.

This variety has singular points. These singular points are the unfurled swallowtails (for a typical obstacle problem, that is for a generic obstacle surface and a generic initial condition). The straight lines corresponding to the cusped edge of the unfurled swallowtail, have an asymptotical direction at the point of tangency with the obstacle surface. The vertex of the unfurled swallowtail is the line, which is in addition tangent to the curve on the obstacle surface formed by the points where the geodesic of our family has an asymptotical direction (fig. 23).

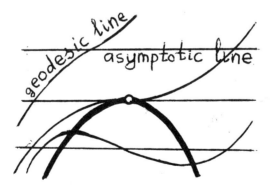

Fig. 23 - The curve of tangency of the geodesic lines with the asymptotic lines.

The unfurled swallowtail is one of the simplest members of a large family of singularities. Let us consider the tower of the fibrations

$$\ldots \to \mathbb{C}^{n+1} \to \mathbb{C}^n \to \ldots \to \mathbb{C},$$

formed by the spaces of the polynomials

(++)
$$\mathbb{C}^n = \{e_{n+1} + \lambda_1 e_{n-1} + \ldots + \lambda_n e_1\}, \quad \text{where}$$

$$e_k = x^k/k!,$$

and by the differentiation operations. Differentiation decreases by one both the degree of the polynomials and the multiplicities of their multiple roots. Hence a polynomial having a root of fixed comultiplicity is projected to a polynomial having a root of the same comultiplicity (the *comultiplicity* being the difference between the degree of a polynomial and the multiplicity of a root).

EXAMPLE. Over the swallowtail $\Sigma_2(4)$, formed by the polynomials (++) of degree 4 having a root of comultiplicity at most 2, there lies at the next floor of our tower the unfurled swallowtail $\Sigma_2(5)$, formed by the polynomials (++) of degree 5 with a root of comultiplicity at most 2. At the next floor of the polynomials of degree 6 lives the surface $\Sigma_2(6)$, which is projected onto $\Sigma_2(5)$ by the differentiation, and so on.

All these surfaces $\Sigma_2(n)$, $n \geq 5$, happen to be diffeomorphic to each other (and hence diffeomorphic to the unfurled swallowtail surface).

Returning to the general case, let us consider the variety of the polynomials (++) of degree d having a root of comultiplicity at most k. The dimension of this variety $\Sigma_k(d)$ is equal to k for any $d \geq k$. We obtain a tower of the algebraic varieties

$$\ldots \to \Sigma_k(d+1) \to \Sigma_k(d) \to \ldots \to \Sigma_k(k+1) = \mathbb{C}^k,$$

where the mappings are differentiations of the polynomials.

This tower stabilizes at the floor of the *k-dimensional unfurled swallowtail* $\Sigma_k(2k+1)$ - the lowest floor where all the self-intersections disappear (A.B. Givental [8]).

All the varieties $\Sigma_k(d)$, with $d \geq 2k+1$, are diffeomorphic to each other.

PROBLEM. Prove that the fibration $\mathbb{C}^{n+1} \to \mathbb{C}^n$, with $n \geq 2k$, admits a smooth section containing the variety $\Sigma_k(n+2)$.

EXAMPLE. The cusped edge $\Sigma_1(4)$ of the ordinary swallowtail lies on a surface which is transversal to the axis λ_3 (fig. 24).

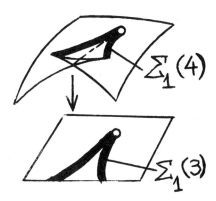

Fig. 24 - The cusped edge of a swallowtail, projected to the 1-dimensional open swallowtail.

HINT. The required hypersurface is a paraboloid

$$\lambda_{n+1} = F(\lambda_1, \ldots, \lambda_n),$$

where F is a quadratic form (known to Hilbert).

Since such a hypersurface is projected diffeomorphically on $\mathbb{C}^n$, the variety $\Sigma_k(n+2)$, lying in this hypersurface, is projected onto $\Sigma_k(n+1)$ diffeomorphically. Hence all the varieties of the polynomials of sufficiently high degree, having a root of comultiplicity (at most) k, are diffeomorphic to each other (and to the k-dimensional unfurled swallowtail $\Sigma_k(2k+1)$).

The role of these varieties in the multidimensional obstacle problem is similar to the role of the 2-dimensional unfurled swallowtail in the 3-dimensional obstacle problem. Namely, they provide the normal forms for the typical singularities of the systems of the extremals of the ambient space, tangent to the extremals of the problem we deal with on the surface of the obstacle.

PROBLEM (O.P. Shcherbak). The space curves, projectively dual to the curves of a generic 1-parameter family containing a curve with a biflattening point ($y = x^2 + \ldots$, $z = x^5 + \ldots$), sweep a surface having an unfurled swallowtail singularity in the space-time.

HINT. The curve, projectively dual to a given curve in projective n-space, is the curve, in the dual projective space, formed by the osculating hyperplanes of the given curve, these hyperplanes being considered as points of the dual projective space.

1.8 - The folded and the open umbrellas

The *folded umbrella* is the surface $y^2 = z^3 x^2$ (fig. 25).

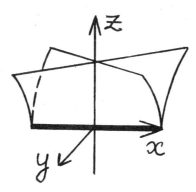

Fig. 25 - The folded umbrella.

It has a smooth semicubical cusped edge and a self-intersection line. Topologically this surface (in real 3-space) has the same structure as the ordinary Cayley umbrella (fig. 7).

EXAMPLE. Let us consider a smooth curve in Euclidean (or projective) 3-space. The curvature and the torsion of a generic curve at a generic point are both different from zero (the curve has a quadratic tangency with the tangent line and a cubical tangency with the osculating plane). But a generic curve may contain some isolated *flattening points*. In a neighbourhood of such a point, the equation of the curve may be written (in suitable affine coordinates) in the form

$$y = x^2 + \ldots, \quad z = x^4 + \ldots,$$

where the dots signify the higher order terms.

Let us consider now the union of all the tangent lines of our curve. This developable surface has a semicubical cusped edge - the initial curve. The surface formed by the tangent lines of a generic curve has folded umbrella singularities at the flattening points of the initial curve (O.P. Shcherbak [9]).

The transition to the curves in higher-dimensional spaces yields higher-dimensional folded umbrellas (in the neighbourhood of the "flattening" point the curve in an n-space admits a representation

$$x_k = x_1^k + \ldots (k < n), \quad x_n = x_1^{n+1} + \ldots .)$$

Folded umbrellas occur in many other problems of singularity theory.

EXAMPLE. Let us consider the folding mapping of the 3-space

$$y_1 = x_1^2, \quad y_2 = x_2, \quad y_3 = x_3.$$

Suppose a surface is given in the x-space having a semicubical cusped edge situated generically with respect to the folding (the edge intersecting transversally the fold surface $x_1 = 0$ and the kernel $\partial/\partial x_1$ of the differential of the folding map should not be tangent to our surface at the intersection point of the edge with the fold surface).

The image to which the surface with a semicubical cusped edge is sent by the folding mapping, has a folded umbrella singularity (at the image of the intersection point of the edge with the fold surface).

Folded umbrellas occur in optics as the typical singularities of the "bicaustics" in 3-space. Let us consider a 1-parameter family of surfaces in Euclidean 3-space. Each surface of the family defines its *caustic* (the set of curvature centres, or the envelope of the normal rays). The caustic of a typical surface has a semicubical cusped edge.

When the parameter ("the time") varies, the caustic moves. The moving cusped edge sweeps a surface, which is called the *bicaustic*. The folded umbrella is one of the typical singularities of the bicaustics. It was in this way it appeared in singularity theory [10].

EXAMPLE. Let us consider a generic projection from a 4-space to a 3-space of a 2-dimensional surface having a semicubical cusped edge (and hence being locally diffeomorphic to the direct product of a line in a plane with a semicubical parabola in another plane).

The tangent plane of such a surface is vertical (contains the direction of projection) at some isolated points of the cusped edge. The image to which the projection sends our surface, is a surface in the 3-space having folded umbrella singularities (at the images of these isolated points).

PROBLEM (O.P. Shcherbak [7]). Prove that the quotient space of the D_4 caustic in $\mathbb{C}^3$ with respect to the reflection in the plane, containing a cusped edge, is diffeomorphic to the folded umbrella.

REMARK. In the real domain, the folding of the pyramid (fig. 11) provides the part of the folded umbrella lying from one side of the image of the mirror of the folding, while the folding of the purse provides the other part (fig. 26).

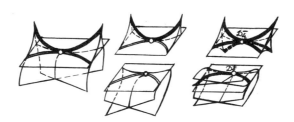

Fig. 26 - The folding of the pyramid and of the purse over the "pyralek".

This observation led O.P. Shcherbak to the name "pyralek" for the union of the folded umbrella with its tangent plane at the vertex ("pyralek" from *"pyramid"* and *"koshelek"*, which is the Russian word for "purse").

The open umbrella is the 2-dimensional surface in the 4-space, with an isolated singular point. This surface is the image of the mapping $(u,v) \mapsto (x,y,z,w)$,

$$x = u^2, \ y = u^3, \ z = v, \ w = uv.$$

A typical mapping of a 2-surface in the 4-space has no singular point. But in the typical one-parameter families of such mappings, the singularities occur for some isolated values of the parameter, and these singularities are (diffeomorphic to) the open umbrellas.

This singular surface may be represented as being swept in space-time by a moving space curve. A typical space curve (a generic mapping of a line into the 3-space) has no singular points. The isolated singular points occur in generic two-parameter families of space curves for some special values of the parameter. The curve, corresponding to such a special value, has a semicubical cusp point.

A typical one-parameter deformation of a space curve having a cusp, is reducible to the normal form

$$x = u^2, \ y = u^3, \ w = uv \text{ (where } v \text{ is the parameter)},$$

by smooth changes of one independent and of the three dependent variables, depending smoothly on the parameter, and by a smooth change of the parameter.

The surface swept by these curves in space-time is the open umbrella. The projection of this surface to the (x,z,w)-space along the y axis is the usual Whitney-Cayley umbrella (fig. 7). The lift from the ordinary umbrella to the open umbrella is the normalization, similar to the lift from the ordinary swallowtail to the open swallowtail.

PROBLEM (A.B. Givental [11]). Prove that the conormal bundle of a semicubical parabola (formed by the linear forms on the tangent spaces of the plane at the points of the semicubical parabola vanishing on the tangent vectors of the parabola) is diffeomorphic to the open umbrella. The standard semicubical parabola here may be replaced by any curve diffeomorphic to it (having a semicubical cusp).

The solution of this problem allows us to define the higher-dimensional open umbrellas. These are even-dimensional subvarieties in spaces of dimension divisible by four, namely the conormal bundle varieties of the higher-dimensional open swallowtails (the semicubical parabola being the 1-dimensional open swallowtail).

PROBLEM. Prove that the $2n$-dimensional open umbrella is diffeomorphic to the set of the values of the parameters (a,b), such that the two polynomials

$$x^{2n+1} + a_1 x^{2n-1} + \ldots + a_{2n}, \ b_1 x^{2n-1} + \ldots + b_{2n}$$

have a common root of multiplicities at least $(n+1, n)$.

PROBLEM (A.B. Givental). Prove that the strata of the singularities of the $2n$-dimensional open umbrella are diffeomorphic to the open umbrellas of lower (even) dimension.

CONJECTURE (A.B. Givental). The set of mappings whose images have no singularities other than the open umbrellas, is open and dense in the space of the isotropic smooth mappings of 2-dimension surfaces to the 4-dimensional space, equipped with a nondegenerate skew-symmetric differential two-form (a mapping is isotropic if the induced two-form on the surface vanishes identically).

Givental [11] has proved that this set of isotropic mappings is open in the set of all isotropic mappings, but not the density.

1.9 - The singularities of the projections and of the apparent contours

Let us consider a generic smooth surface in the usual 3-space. The shape we see depends on the point of view. What we see is the apparent contour of the projection to the plane (to the retina surface). For a generic point of view, the only critical points of the projection are the Whitney folds and cusps (1.2). Consequently, an apparent contour has generically no singularities besides the semicubical cusp points (and the transversal self-intersection points).

But seen from some special viewpoints, the surface may have other apparent contour singularities. In a generic 3-parameter family of mappings, singularities more complicated than the Whitney folds and tucks may occur (the three parameters being the three coordinates of the point of view).

The number of nonequivalent singularities of the projections of generic surfaces in the usual 3-space, defined by the families of rays issued from different points of space outside the surface, is finite; in fact it is equal to 14 (O.A. Platonova, O.P. Shcherbak, 1981, see [12] and [7]). The hierarchy of the adjacencies of these singularities is

$$
\begin{array}{ccccccccc}
1 & \leftarrow & 2 & \leftarrow & 3 & \leftarrow & 6 & \leftarrow & 8 \\
 & & & & \uparrow & & \uparrow & & \uparrow \\
 & & & & 4 & \leftarrow & 7 & & 11 \\
 & & & & \uparrow & & \uparrow & & \\
 & & & & 5 & \leftarrow & 10 & & \\
 & & & & \uparrow & & & & \\
 & & & & 9 & & & &
\end{array}
$$

The numbers here denote the projections of the parts of the surfaces $z = f(x, y)$ (close to origin) by the rays parallel to the x axis. In the nonsingular case 1, the surface is defined by the function $f = x$. The other functions f are given by the following table

type	$f(x,y)$	type	$f(x,y)$
2	x^2	7	$x^4 + x^2y + xy^2$
3	$x^3 + xy$	8	$x^5 \pm x^3y + xy$
4	$x^3 \pm xy^2$	9	$x^3 \pm xy^4$
5	$x^3 + xy^3$	10	$x^4 + x^2y + xy^3$
6	$x^4 + xy$	11	$x^5 + xy$

Thus, 1 is a nonsingular mapping (a diffeomorphism), 2 is a fold mapping, 3 is a Whitney tuck. The 4 and 6 singularities have codimension 1 (are visible from the points of some surfaces), 5, 7, 8 have codimension 2 (are visible from the points of some curves), 9, 10, 11 have codimension 3 (are visible from some isolated points of the space).

A singularity may become simpler under a small variation of the point of view, as it is indicated by the arrows in the hierarchy diagram.

The above classification defines the projection mapping locally up to an equivalence. Two projections of the surfaces are *equivalent* if one may be transformed into the other by a local diffeomorphism of the ambient space sending the first surface to the second one, and the family of rays defining the first projection into the family of rays defining the second one.

The apparent contours of the equivalent projections are diffeomorphic plane curves. The curves corresponding to the 13 projections of the above list, are shown in fig. 27.

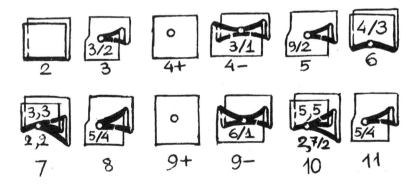

Fig. 27 - *The singularities of the apparent contours of surfaces.*

THE THEORY OF SINGULARITIES AND ITS APPLICATIONS

The finiteness of the list of the typical singularities of the projection mappings is by no means evident *a priori*; rather it is even surprising. Indeed, in the generic 3-parameter families of mappings between surfaces, a continuum of locally nonequivalent mappings occurs. The list of nonequivalent projection mappings is finite only because the 3-parameter family of the projections of the same generic surface from the different centres is a very special, not general, family (mainly due to the strong connection between the projections from the centres lying on the same ray).

Suppose we look at a generic surface from one of the most singular points (9, 10, or 11). Slightly moving the head, we shall see the patterns shown in the fig. 28-31.

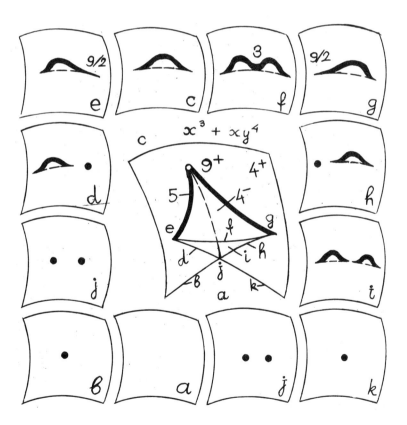

Fig. 28 - The perestroikas of the projection of a surface at the singularity 9^+.

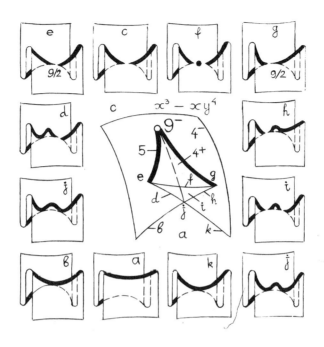

Fig. 29 - *The perestroikas of the projections of a surface at the singularity* 9^-.

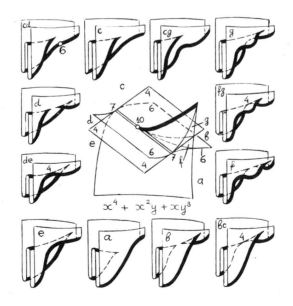

Fig. 30 - *The perestroikas of the projections of a surface at the singularity* 10.

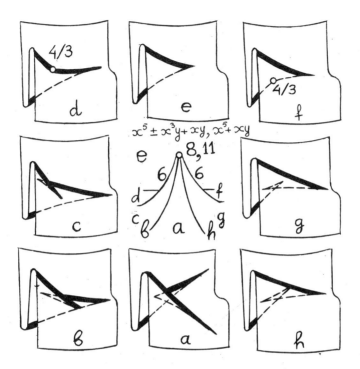

Fig. 31 - The perestroikas of the projections of a surface at the singularities 8 and 11.

In the middle of each figure the bifurcation diagram is represented, formed by the nongeneric centres of the projection. The domains into which the bifurcation diagram cuts the 3-space are denoted by the letters, shown also in the pictures, representing the apparent contours and their perestroikas occuring at the transition through the bifurcation diagram in different places.

2. - SINGULARITIES OF BIFURCATION DIAGRAMS

At first glance, the variety of different singularities encountered in different applications produces an impression of a chaos. A remarkable discovery of the last decades in this domain is of simple general laws governing very different phenomena. The coincidence of the answers in apparently different problems is not an accident: there exist many hidden relations between apparently unrelated objects in singularity theory.

In most of these problems, the object of interest is the bifurcation diagram formed by the values of the parameters for which a "qualitative change" in the objects of the family occurs. The objects forming the families may be very different: manifolds or mappings, vector fields or differential equations, abelian differentials or integrals, and so on.

In a real space of the parameters, a bifurcation diagram is mostly a hypersurface cutting the parameter space into parts formed by the objects which are "qualitatively the same". A complex hypersurface does not cut the complex parameter space into parts. In the complex case, the separating role of the real bifurcation diagram is transformed into that of the branching set for the monodromy (which is the natural action on our objects of the fundamental group of the complement to the bifurcation diagram).

In this part we consider some examples of bifurcation diagrams related (in the real case) to the boundaries of some natural domains in the functional spaces of differential equations: the domains of stability, of ellipticity, of hyperbolicity, and so on.

2.1 - Bifurcation diagrams of families of functions

Let us consider the subset of functions having a critical value zero in the space of all smooth functions. This subset is called the *bifurcation set of zeroes*. Indeed, the space of functions may be considered as a (linear) surrogate of the space of hypersurfaces (each function representing its zero level hypersurface). From this point of view, the bifurcation set of zeroes is the subset of singular hypersurfaces in the space of all the hypersurfaces.

The bifurcation set of zeroes has a natural stratification (different strata corresponding to functions having different numbers of critical points of different types). The hierarchy of the critical points (of the holomorphic functions of $n \geq 2$ variables) is represented in fig. 32.

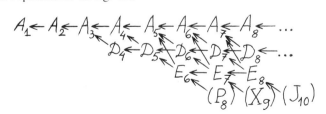

Fig. 32 - The adjacency diagram of the simple singularities of functions.

The letters in this figure denote the strata consisting of the functions which are locally equivalent to the following normal forms:

$$A_k \quad x_1^{k+1} + x_2^2, \ k \geq 1,$$
$$D_k \quad x_1^2 x_2 + x_2^{k-1}, \ k \geq 4,$$
$$E_6 \quad x_1^3 + x_2^4,$$
$$E_7 \quad x_1^3 + x_1 x_2^3,$$
$$E_8 \quad x_1^3 + x_2^5.$$

The function is *equivalent* to the normal form if it can be reduced by a biholomorphic change of independent variables (in a neighbourhood of the critical point) to the sum of the normal form, of a constant and of the nondegenerate quadratic form $x_3^2 + \ldots + x_n^2$ (if the number of variables is $n > 2$).

The arrows indicate the adjacencies of the smaller strata (strata of higher codimension, consisting of more complicated singularities) to the larger strata (consisting of less complicated singularities). A small variation of a function may transform its critical point into one or into several critical points of those types to which there leads a path formed by the arrows from the stratum of the given critical point. The codimension of a stratum is smaller by one than the index in the notation (which is equal to the multiplicity, that is to the number of the Morse critical points of the type A_1, colliding at the critical point). The strata of equal codimension are shown in fig. 32 at the same vertical column. The three letters in brackets are not strata, but three sets, containing all the singularities more complicated than those on the list. They are all adjacent to E_6, E_7 or E_8 and they form a set of codimension 6 in the space of all functions of $n > 2$ variables (7 for $n = 2$, ∞ for $n = 1$).

Thus, a typical k-parameter family of functions of n variables, depending on 5 or fewer parameters, contains no singularity other than the A_ℓ, D_ℓ, E_ℓ singularities (with $\ell \leq k + 1$).

The A, D, E singularities are characterized by their property of being *simple*: the set of the nonequivalent singularities, generated by their small variations, is *finite* for each of these singularities.

All the other singularities have continuous invariants (moduli).

EXAMPLE. The simplest nonsimple singularity P_8 is defined by the nondegenerate cubical form in three variables

$$f = x_1^3 + x_2^3 + x_3^3 + ax_1 x_2 x_3.$$

The singularities corresponding to the different values of the parameter a, are generically nonequivalent (since the corresponding elliptical curves $f = 0$ in $\mathbb{C}P^2$ cannot be holomorphically transformed one into the other).

The trace of the bifurcation set of the zeroes on the parameter space of a family of functions is called the *bifurcation diagram of the zeroes of the family*.

EXAMPLE. The bifurcation diagram of zeroes of the family (of functions of x, depending on the parameters (λ_1, λ_2)):

$$f(x, \lambda) = x^3 + \lambda_1 x + \lambda_2,$$

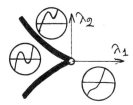

Fig. 33 - The bifurcation diagram of zeroes of singularity A_2.

is the semicubical parabola $\Delta = 0$, where $\Delta = 4\lambda_1^3 + 27\lambda_2^2$ is the discriminant (fig. 33).

A generic family intersects the strata of simple singularities (A, D, E) transversally. The singularity of the bifurcation diagram of zeroes of such a family, at every point of a simple stratum, is the same (up to a biholomorphic diffeomorphism and up to multiplication by a smooth space, in the case where the number of the parameters is greater then the codimension of the stratum). In particular, the bifurcation diagram is independent of the number n of the arguments of the functions in our family (which number, hence, might be even infinite) and on the choice of a generic family.

THEOREM. *The bifurcation diagram of zeroes of a generic family of functions is biholomorphically equivalent to the discriminant of the corresponding finite reflection group (multiplied by a smooth space if the number of parameters is greater then the codimension of a stratum) at a neighbourhood of each point of each simple singularity stratum A, D, E (see 1.5 above and fig. 32).*

This theorem is astonishing, since there is no evident relation between the classifications of the Weyl groups of simple Lie algebras and of the simple singularities of functions. The proofs are based on a comparison of the two independent classifications and on a more or less direct comparison of the discriminants with the bifurcation diagrams. The coincidence confirms the general principle of unity of all mathematical beings (like the relation between the problems on tangents and on areas, basic in calculus).

The list of simple Lie algebras contains, besides the $(A, D\ E)$ algebras which have only simple edges in their Dynkin diagrams, also those algebras having a double edge (B_k, C_k, F_4) and the algebra G_2 with a triple edge.

The algebras having a double edge correspond to the critical points of functions on manifolds with a boundary. Their hierarchy starts with the graph of fig. 34.

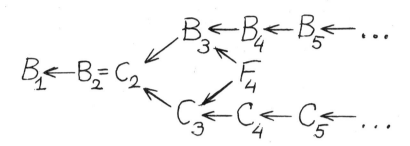

Fig. 34 - The adjacency diagram of simple boundary singularities.

This graph contains all the simple singularities. The index denotes the number of colliding singularities and is equal to the codimension of the stratum in the space of functions. The normal forms are

$$B_k \quad x^k + y^2, \ k \geq 2;$$
$$C_k \quad xy + y^k, \ k \geq 3;$$
$$F_4 \quad x^2 + y^3.$$

A function having a critical point, with zero critical value at a point of the "boundary" $x = 0$ of the plane $\{(x, y)\}$, is reducible to this normal form by a biholomorphic diffeomorphism of the plane which preserves the "boundary" line.

In the case of functions on $n > 2$ variables, one should add the nondegenerate quadratic form $y_2^2 + \ldots + y_{n-1}^2$ to the normal form. In the real case, the monomials entering in the normal forms should be equipped with signs. All the other (nonsimple) singularities form a set of codimension 5 in the space of functions on a manifold with a boundary, and they are adjacent either to F_4 or to C_4.

Hence a typical family of functions on a manifold with a boundary depending on 4 or fewer parameters, does not contain any function having a nonsimple critical point.

The bifurcation set of zeroes of functions on a manifold with a boundary consists of those functions, whose zero level hypersurfaces are *either* singular (A_1) *or* nontransversal to the boundary (B_1). Hence, this set consists of *two* hypersurfaces.

Consequently, its trace on the parameter space of a typical family (called the bifurcation diagram of zeroes of the family) consists of two hypersurfaces.

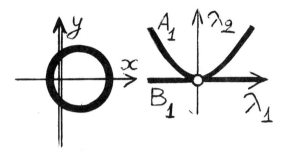

Fig. 35 - *The bifurcation diagram of zeroes of the boundary singularity B_2.*

EXAMPLE. Let us consider the family

$$f(x, y; \lambda) = y^2 + x^2 + \lambda_1 x + \lambda_2$$

of functions on the plane $\{(x, y)\}$ with a "boundary" $x = 0$, depending on the parameters (λ_1, λ_2). The bifurcation diagram of zeroes of the family (fig. 35) consists on those values of the parameters for which the circle $f = 0$ either degenerates (the A_1 component, $\lambda_1^2 = 4\lambda_2$) or is tangent to the axis of y (the B_1 component, $\lambda_2 = 0$).

The singularity at $\lambda = 0$ is of type B_2.

THEOREM. *The bifurcation diagram of zeroes of a generic family of functions on a manifold with a boundary, is biholomorphically equivalent to the discriminant of the corresponding reflection group at the points corresponding to a simple boundary singularity whose codimension is equal to the number of the parameters (if the number of the parameters is greater by d, the discriminant should be multiplied by a smooth space of dimension d).*

PROBLEM. Which of the two branches of the discriminant of the group $B_3 = C_3$ (shown in fig. 15) corresponds to the singular level hypersurfaces, and which to those nontransversal to the boundary?

Hint. It is sufficient to consider the families of functions

$$f = y^2 + x^3 + \lambda_1 x^2 + \lambda_2 x + \lambda_3 \quad \text{for } B_3,$$
$$f = xy + y^3 + \lambda_1 y^2 + \lambda_2 y + \lambda_3 \quad \text{for } C_3.$$

ANSWER. For B_3, the smooth component corresponds to nontransversality, and the singular component to degeneration; for C_3, the roles are reversed (fig. 36).

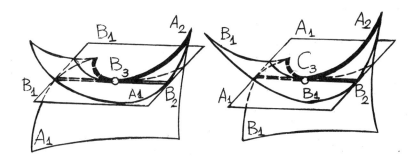

Fig. 36 - The bifurcation diagrams of zeroes of the boundary singularities B_3 and C_3.

REMARK 1. The duality between the B_k and C_k singularities, whose bifurcation diagrams differ only in the roles of components, is a particular case of the general Lagrange duality, which is the duality between the function $f(x,y)$ on a manifold $\{(x,y)\}$ with a "boundary" $x = 0$ and the function $F = f(x,y) + xz$ on a manifold $\{(x,y,z)\}$ with a "boundary" $z = 0$ (where z is the "Lagrange multiplier"). This duality exchanges the types of a function on the ambient space with the type of its restriction to the boundary (F is equivalent to the restriction of f to the boundary, the restriction of F to the boundary being f). For more details, see the article of I.G. Shcherbak [13].

REMARK 2. The theory of the boundary singularities is equivalent to the theory of singularities of the functions, even with respect to one of the variables. Indeed, a function, even in X, may be viewed as a function depending on $x = X^2$ on a manifold whose boundary is $x = 0$. From this point of view the missing Lie algebra G_2 is also included in the general setting: it corresponds to a simple singularity of a function, which is invariant under the action of the symmetry group of a triangle.

REMARK 3. The crystallographic character of the Weyl groups of the simple Lie algebras has the following meaning in terms of the singularities. The Euclidean space where the reflection group acts, is interpreted as being the real middle dimension ($n - 1 = 4s$) homology space of a nonsingular level hypersurface of the function, equipped with the intersection form. The crystallographic lattice is formed by the integer homology classes. The reflection group is the monodromy group (the natural representation of the fundamental group of the complement to the complex bifurcation diagram of zeroes by the automorphisms of the homology space of the nonsingular level hypersurface).

In the case of boundary singularities, we first consider the double branched covering ($x = X^2$), and then we consider the "Prym homology" space, consisting of homology classes changing their signs under the involution of the covering, changing the sign of X. The details may be found in [14].

The simple singularities may be characterized in this theory as the *elliptical* singularities: they are the only singularities having a positive definite intersection form.

Besides the bifurcation diagrams of zeroes, one frequently encounters the bifurcation diagrams of *functions*. The space of functions contains, with every function, its sums with different constants. These functions have different zero level hypersurfaces, but their critical point types are independent of the constant we add. Hence the bifurcation diagrams of functions (the caustics and the Maxwell sets) are cylindrical (invariant under the addition of a constant to all the functions).

EXAMPLE. Let us consider the 3-parameter family of complex functions

$$f = x^4 + \lambda_1 x^2 + \lambda_2 x + \lambda_3$$

(fig. 37).

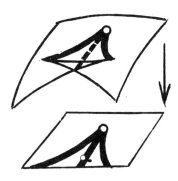

Fig. 37 - *The bifurcation diagram of functions of the singularity A_3.*

The bifurcation diagram of zeroes is a swallowtail surface. The caustic of the family is a cylinder, whose generator is vertical (parallel to the axis of λ_3). The Maxwell set is a vertical plane. The bifurcation diagram of functions of the family is the union of a semicubical parabola in the plane (λ_1, λ_2) with its tangent line at the origin, multiplied by the vertical line.

Any generic projection of the 3-space to a plane may be locally tranformed into the projection along the axis λ_3 by a swallowtail preserving biholomorphic local diffeomorphism. Moreover, any generic vector field may be transformed by such a diffeomorphism into the standard vertical field $\partial/\partial \lambda_3$.

Similar results hold for all the simple singularities.

THEOREM. *The bifurcation diagram of functions of a typical family of functions, containing a function with a simple singularity, is locally biholomorphically equivalent to the projection of the set of the singular points of the discriminant hypersurface of the corresponding reflection group along the fibres of any one-dimensional fibration transversal to the discriminant (the projection should be multiplied by a smooth space if the number of the parameters exceeds the codimension of the singularity).*

The projection in this theorem may be always chosen to be the projection of the space of invariants along the axis of the invariant of the highest degree.

The caustic is the projection of the cusped edge of the discriminant (of the closure of the stratum A_2). The Maxwell set is the projection of the set $2A_1$ of selfintersections of the discriminant.

The caustics of the boundary singularities have generically 3 components: they are the projections of the cusped edges of the two components (A_1 and B_1) of the discriminant and the projection of the intersection of these components.

EXAMPLE. The caustic of the singularity F_4 is shown in fig. 38.

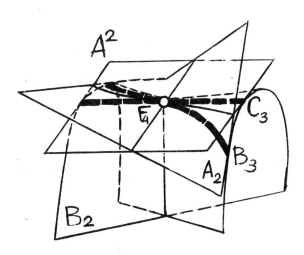

Fig. 38 - The caustic of the reflection group F_4.

Let us consider a generic surface with a generic boundary in Euclidean 3-space. The boundary may be tangent to the curvature lines of the surface at some isolated points of the boundary. The focal set of a surface with boundary consists of the 3 parts:

1) the surface curvature centre set (the envelope of the normal lines).

2) the boundary curve curvature centre set (the envelope of its normal planes).

3) the union of the normals of the surface at the points of the boundary curve.

The union of these three surfaces is locally diffeomorphic to the caustic of F_4 (at a neighbourhood of the curvature centre on the normal to the surface at the point of tangency of the boundary with the curvature line), the surface and the boundary being generic (I.G. Shcherbak [15]).

2.2 - Stability boundary

A system of linear differential equation $\dot{x} = Ax$ is *stable*, if the real parts of the eigenvalues of the operator A are all negative. The stable systems form the *stability domain* in the space of linear operators A.

A family of linear operators is a mapping from the parameter space to the space of linear operators. The *stability domain of a family* is the pre-image of the universal stability domain. It is a domain of the parameter space. Its boundary is the *stability boundary* of the family.

EXAMPLE. The stability domain of the two-parameter family of the linear systems $\ddot{x} + a\dot{x} + bx = 0$ is the quadrant $a > 0$, $b > 0$.

The stability boundaries of the typical two-parameter families of the linear systems have no other singularities than the angles between two transversal branches of the boundary.

THEOREM (L.V. Levantovskii [16]). *The nondiffeomorphic singularities of the stability boundary, occuring in typical k-parameter families of linear systems, form a finite set for any fixed number of the parameters k and this set is independent of the order of the system, if it is greater then some $n(k)$.*

EXAMPLE. The only singularities of the stability boundaries occuring in typical three-parameter families of linear systems are (see fig. 39):

1)   the two-face angle ($a \geq 0$, $b \geq 0$);

2)   the three-face angle ($a \geq 0$, $b \geq 0$, $c \geq 0$);

3)   the Whitney half-umbrella ($b^2 \leq ca^2$, $a \geq 0$);

4)   the flattened angle ($b^2 \leq c^2a^2$, $a \leq 0$, $c \leq 0$).

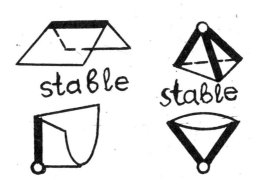

Fig. 39 - The typical singularities of the stability boundary in the parameter 3-space.

The above formulae provide the normal forms of the stability boundary together with the stability domain in the space of the parameters (a, b, c).

The stability domain forms the smaller of the two domains, bounded by the stability boundary (in all 4 cases). This property (holding also for the typical families of linear systems depending on an arbitrary number of the parameters) is a manifestation of a very general "principle of fragility", according to which a small displacement from a singular point of the boundary of a domain of "good objects" makes the object more likely to be bad than good.

The smaller domain may be asymptotically infinitely smaller (like in the case 4 above). Singularities of this type on the boundary are particularly dangerous for such systems, like atomic reactors, which always work at the stability boundary by their very nature.

The typical singularities of the stability domains in k-parameter families of systems with n-dimensional phase spaces, stabilize for growing n (the list of singularities becomes independent of the dimension of the phase space, that is of the order of the matrix A). Hence the singularities are the same even for the infinite dimensional systems defined by the families of linear operators (of a suitable class) in a space of infinite dimension.

2.3 - Ellipticity boundary and minima functions

A linear differential operator is *elliptic* if its principal symbol is positive definite. The *ellipticity domain* of a family of operators is defined as the set of values of the parameters corresponding to the elliptic operators. The boundary of this domain is called the *ellipticity boundary.*

Let us consider k-parameter families of the differential operators of order d acting on functions on n variables. Let us fix the dimension k of the parameter space.

THEOREM (V.I. Matov [17]). *The singularities of the ellipticity boundaries in the typical k-parameter families stabilize in the following 3 cases:*

a) $n \to \infty$ *(the stable singularities then depend on d);*

b) $d \to \infty$ *(the stable singularities then depend on n);*

c) $n \to \infty$, $d \to \infty$ *(the bistable singularities depend only on k).*

The bistable $(n \geq n_0(k), d \geq d_0(k))$ *singularities of the ellipticity boundaries (and of the ellipticity domains), in the typical families, coincide with the singularities of the graphs (and of the subgraphs) of the minima functions*

$$F(\lambda) = \min\ f(x, \lambda)$$

of the typical (k-1)-parameter families of functions on the compact manifolds, whose dimensions are sufficiently high (these singularities also stabilize when the dimension is growing). The last singularities are also diffeomorphic to the singularities of the level hypersurfaces (and of the sets where the function is negative) of the minima functions of the typical k-parameter systems.

If the dimension of the parameter space is smaller then 7, the number of nondiffeomorphic singularities of the graphs of the minima functions in typical families is finite, and all the singularities occur as the singularities of minima functions of the generic families of the polynomials in one variable. The hierarchy of the possible singularity types is shown in fig. 40 (where A_1 corresponds to a nonsingular point of the graph).

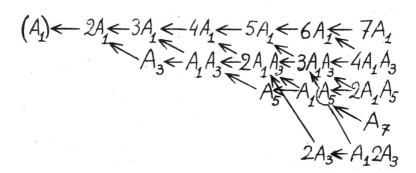

Fig. 40 - The adjacency diagram of the minima function singularities.

The singularity (pA_1) (qA_3) (rA_5) ... first occurs in the typical $(p + 3q + 5r + \ldots - 1)$-parameter families. In the typical two parameter families

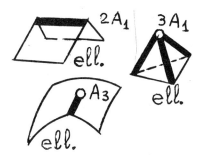

Fig. 41 - The typical singularities of the ellipticity boundary in the parameter 3-space.

only the following three singularities of the graphs of the minima functions occur (fig. 41): the 2-face angle $2A_1$, the 3-face angle $3A_1$, and the "mutilated swallowtail" A_3 (the last one is obtained from the usual swallowtail by the amputation of the pyramid formed by the real root polynomials). From 7 parameters on, the nondiffeomorphic singularities form continuous sets (the moduli occur - first a finite number of them, then the arbitrary functions become moduli).

The "good" set (the ellipticity domain, the subgraph, the negative value domain) always propagates its wedges inside the "bad" domain, thus confirming the fragility principle (holding here for any dimension of the parameter space, since the set of the positive definite forms is convex).

The set of the values of the parameters for which the minimum is attained at more than one point (counted with multiplicities) is called the *Maxwell set of the family* (the minima function is nonsmooth there). The singularities of the Maxwell sets of typical families in the three-dimensional parameter space are shown in fig. 42.

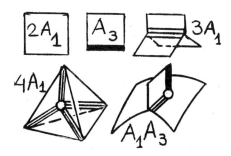

Fig. 42 - The typical singularity varieties of minima functions of three variables.

2.4 - Hyperbolicity boundary

A real projective algebraic hypersurface is *hyperbolic* (with respect to a given point), if all the points of intersection of this hypersurface with any real straight line containing this point are real (fig. 43).

Fig. 43 - A hyperbolical curve of degree four.

EXAMPLE. The light hypersurface (the set of the zeroes of the principal symbol) of the wave equation is hyperbolic with respect to any time-like direction (this "direction" being considered as a point of the projective space of the space cotangent to space-time).

The *hyperbolicity domain* in the parameter space of a family of differential equations is defined as the set of those values of the parameter for which the principal symbol of the equation is hyperbolic.

THEOREM (A.D. Weinstein, B.Z. Shapiro [18]). *The singularities of the hyperbolicity domains of typical k-parameter families stabilize like the singularities of the ellipticity domains when the number of the independent variables or the order of the equation increases. The bistable singularities of the hyperbolicity domains of the typical k-parameter families coincide with the bistable singularities of the ellipticity domains.*

The relation between stable hyperbolicity and stable ellipticity becomes transparent in the theory of the *local hyperbolicity*, introduced by Atiyah, Bott and Garding [19].

A real algebraic hypersurface is *hyperbolic at a point* (with respect to a line element), if any neighbouring real straight line intersects the hypersurface in a neighbourhood of this point only in the real points. In this case any neighbouring real smooth curve intersects the hypersurface (near this point) in the same number of points (counted with the multiplicities). This number, called the *multiplicity*, replaces, in the local case, the degree of the hypersurface, counting intersections in the global case.

Local hyperbolicity is invariant under diffeormophisms (and it may be defined for smooth nonanalytic hypersurfaces).

EXAMPLE. A plane analytic curve is locally hyperbolic (at a singular point), if and only if all its branches at that point are smooth.

The stabilization theorems hold for local hyperbolicity as well as in the global case. However, as the multiplicity of a hyperbolical curve is an arbitrary number (as in the above example), the structure of the generic locally hyperbolical surfaces is much simpler.

THEOREM [18] *The stable locally hyperbolic singularities of hypersurfaces of dimension greater then one are isolated and their multiplicity is equal to two. They are defined by equations of the form* $x^2 = f(y_1, \ldots, y_{n-1})$, *where* $f(y) > 0$, $f(0) = 0$.

Thus, the hyperbolicity condition means that f has a minimum at the origin. This explains why the typical singularities of the boundaries of the hyperbolicity and ellipticity coincide.

Different problems in singularity theory, related to the notion of hyperbolicity, are discussed in [20] - [27].

2.5 - Disconjugate equations, Tchebyshev system boundaries and Schubert singularities in complete flag manifolds

A linear ordinary differential equation of order n is called *disconjugate* if each of its nonzero solutions vanishes at most at $n - 1$ points.

EXAMPLE. The equation $\ddot{x} = x$ is disconjugate. The equation $\ddot{x} + x = 0$ is disconjugate on any segment of length less than π.

A fundamental system of solutions of a disconjugate equation is called a *Tchebyshev system*. Such a system generates a linear space of solutions. Any n-dimensional linear space of functions contains nonzero functions, vanishing at any given set of $n - 1$ points. Tchebyshev systems generate n-dimensional spaces whose nonzero functions never vanish at more than $n - 1$ points.

Let us consider the manifold of *complete flags* $\{V_0 \subset V_1 \subset \ldots \subset V_n\}$, where V_k is a k-dimensional subspace of a fixed n-dimensional linear space.

Let us fix one of the flags. The whole manifold of flags is the union of the submanifolds of flags, situated differently with respect to the fixed flag. A *Schubert cell* is the set of all flags for which the dimensions of the intersections of the spaces of the flags with the spaces of the fixed flag, have the prescribed values. These cells are diffeomorphic to the linear spaces. The whole flag manifold is the union of nonintersecting Schubert cells.

One of the cells is formed by the flags transversal to the fixed one. This *open cell* has the largest dimension (equal to the dimension of the flag manifold). Its complement is a hypersurface, called the *train* of the fixed flag (as the train of a dress). It is the union of all the other (nonopen) Schubert cells.

The singularities of the train along any cell are locally diffeomorphic. The singularities of the boundaries of the disconjugacy domains in the generic k-parameter families of linear differential equations (or the singularities of the boundaries of the Tchebyshev domains of generic k-parameter families of systems of functions) coincide with the singularities of the train of a flag (at the cells of codimension at most k). These singularities stabilize when n increases (B.Z. and M.Z. Shapiro [28]).

The Schubert cells in the space of n-flags (and hence the train singularities) are naturally numbered by the permutations of n symbols $(1,\ldots,n)$. Let us choose a basis such that spaces of the fixed flag are generated by the first; the first and second; the first, the second and the third (and so on) basis vectors. A permutation acting on the basis transforms the fixed flag into a new one. This new flag is contained in a Schubert cell which is uniquely defined by the permutation. All the Schubert cells are provided by this construction.

EXAMPLE. The hierarchy of the Schubert cells (the "Bruhat ordering") in the space of the complete flags in a linear 3-dimensional space is shown in fig. 44.

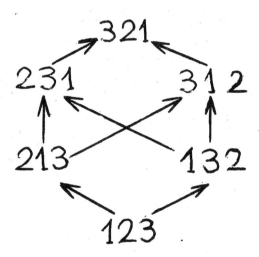

Fig. 44 - *The Bruhat ordering of the Schubert cells in the flag manifold of the 3-space.*

The initial fixed flag forms the 0-dimensional cell $(1, 2, 3)$. The higher-dimensional cells form the higher horizontal lines in fig. 44. The arrows show the direct adjacencies of the cells (to the cells whose dimension is greater by one). Each permutation is related, by the adjacency arrows, with those permutations which are obtained from it either by one transposition of two symbols in neighbouring positions, or by one transposition of symbols differing by one. The dimension of a cell is equal to the number of the disorders in the permutation, comparing it with $(1,\ldots,n)$; the codimension is equal to the number of the disorders comparing with $(n,\ldots,1)$.

The singularity of the train of our fixed flag at its most singular point (1, 2, 3) is diffeomorphic to the surface $z^2 = x^2y^2$, shown in fig. 45.

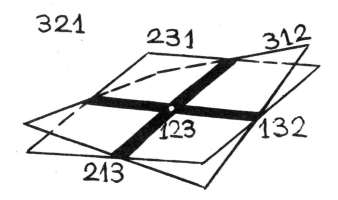

Fig. 45 - The singularities of the train in the flag manifold of the 3-space.

The train cuts the neighbourhood of the most singular point into 6 parts. Four of them are diffeomorphic to the flattened angle of fig. 39.

The typical singularity of the boundary of the disconjugate equations (or of the Tchebyshev systems) are diffeomorphic just to the flattened angle $(x \geq 0, \ y \geq 0, \ z^2 \leq x^2y^2)$.

The $n - 1$ components of the train of a complete flag, in a linear n-dimensional space, define 2^{n-1} domains with boundary in the neighbourhood of this flag, diffeomorphic to each other and to a typical singularity of the domain of disconjugate equations.

PROBLEM. Find the total number of components into which the neighbourhood is cut by the train.

The answer is known only for the few first values of n(4, 6, 20, ?).

The theory of disconjugate equations and of Tchebyshev systems is mainly the geometry of the trains in the space of flags.

For any linear equation of order n, let us consider the spaces of its solutions that vanish at a given point with multiplicity at least $n, \ n - 1, \ldots, 1, \ 0$. These subspaces form a complete flag in the n-dimensional space of all solutions. This flag depends on a point (of the time axis). Thus the equation defines a curve in the space of flags.

PROBLEM (B.Z. and M.Z. Shapiro, [28]). Prove that the sum of the multiplicities of the intersections of a curve defined by a disconjugate equation, with the train of any fixed flag, does not exceed $(n^3 - n)/6$.

PROBLEM (a generalized Sturm theorem, [28]). Suppose a segment of a curve defined by an equation of order n, intersects the k-th component of the train of a fixed flag with a total multiplicity larger than $k(n - k)$. Then this segment intersects the trains of all flags.

Here the k-th component of the train is the variety of flags nontransversal to the k-dimensional space of the fixed flag.

CONJECTURE ([28]). A curve defined by a linear equation of order n, intersecting a train of a flag, passes, at any intersection point, from one of the components, into which a neighbourhood of the intersection point is cut by the train, to a different component.

2.6 - Fundamental system boundaries, projective curve flattenings and Schubert singularities in Grassmann manifolds

A sequence of n functions of one variable is a *fundamental system* of solutions of a linear differential equation of n-th order, if their Wronskian does not vanish at any point. Such sequences form a domain in the functional space of all sequences of n functions. What about the boundary of this domain?

Let us consider the strata of the boundary having codimension at most k. In studying them, we may restrict ourselves to the typical singularities of the bifurcation diagrams in the k-dimensional spaces of the parameters. These singularities stabilize, as is usual, if k is fixed and n is growing. As it is usual, it is simpler to study the analytical continuations of these bifurcation diagrams formed by those values of the parameters for which the Wronskian has a multiple root (whether they belong to the boundary or not).

The results of this section are due to M.E. Kazarijan ([29], [30]).

The only singularity, in two-parameter families, is the cubical tangency of two smooth curves in the plane of the parameters. In the three-dimensional parameter space, three singularities occur:

1) the swallowtail;

2) the folded umbrella with a smooth surface cubically tangent to it (fig. 46);

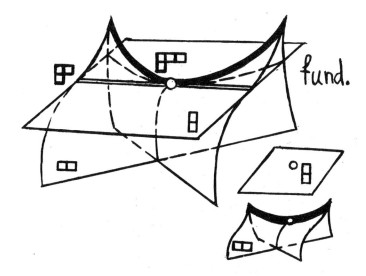

Fig. 46 - A typical singularity of the boundary of the domain of fundamental systems.

3) two Whitney umbrellas, tangent cubically to each other (fig. 47).

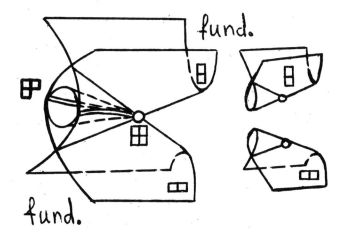

Fig. 47 - A pair of cubically tangent Whitney umbrellas.

Of course, the transversal intersections of surfaces of these types are also possible.

The domains formed by the fundamental systems, are also shown in fig. 46 and 47 (in the swallowtail case this domain is formed by the polynomial with no real roots).

The three surfaces, described above, may be defined as the sets in the space with coordinates (a, b, c), formed by the polynomials with a multiple root of the form:

1) $$t^4 + ut^2 + bt + c$$

2) $$t^4 + at^3 + bt^2 + c$$

3) $$t^4 + 3(a+b)t^2 + ct - 3ab.$$

REMARK. Geometrically, the surfaces of fig. 46 and 47 may be described as the pre-images of the swallowtail surface for the following mappings from a three-space to the three-space containing the swallowtail surface.

Let us consider the cusped edge of the swallowtail surface and choose a smooth surface containing this cusped edge (such surfaces exist and are all locally reducible each to other by swallowtail preserving diffeomorphisms fig. 48).

Fig. 48 - The two-fold covering (over the space containing the swallowtail) branching surface.

The desired mapping is a folding whose set of critical values coincides with the surface we have chosen.

PROBLEM. Prove that the pre-image of the swallowtail looks like the surface shown in fig. 47.

To obtain the folded umbrella surface, we start with a semicubical cylinder (any surface with a semicubical cusped edge passing through the vertex of the swallowtail transversally to the swallowtail surface and containing the cusped edge of the swallowtail, fig. 49). This cylinder is the set of critical values of a Whitney "tuck" mapping from a three-space to our three-space.

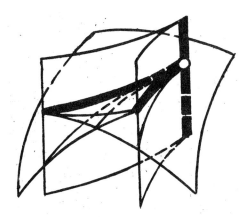

Fig. 49 - The critical value surface of the Whitney tuck mapping to the space, containing the swallowtail.

PROBLEM. Prove that the pre-image of the swallowtail looks like the two umbrellas surface shown in fig. 46.

The union of the folded umbrella with a smooth surface occurs also in other branches of the singularity theory (see fig. 26). The pair of umbrellas occurs in the first problem of 1.3. Let us consider a typical perestroika of a plane curve depending on a parameter and having a semicubical cusp point for some value of the parameter. We put each curve in a horizontal plane at a height equal to the value of the parameter. The curves sweep an umbrella surface. Their inflectional tangents sweep another umbrella surface (fig. 50).

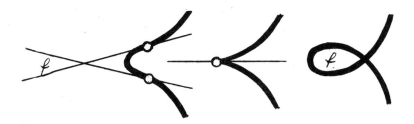

Fig. 50 - The pair of the umbrellas as the set of singular centres of projections of a plane curve depending on a parameter.

The two umbrellas together form the surface of fig. 47.

To describe the relation of our problem to Schubert cells, let us consider the Grassmann manifold of all n-dimensional planes in an N-dimensional linear space. We choose a complete flag in this space. A *Schubert cell* is formed by those n-planes whose intersections with the spaces of the flag have fixed dimensions.

The typical planes form an *open cell*. The union of the other cells is called the *train* of the smallest (0-dimensional) cell. The train is a hypersurface, stratified into cells. The trace of the train on a k-submanifold, transversal to a cell of codimension k in the Grassmann manifold, is (up to a local diffeomorphism) independent of the submanifold and of the point of its intersection with the fixed cell. Moreover, these traces stabilize when n and N are growing while k remains fixed.

EXAMPLE. The stratification contains one cell of codimension 1 and two cells of codimension 2 whose closures intersect along a cell of codimension 3 (fig. 51).

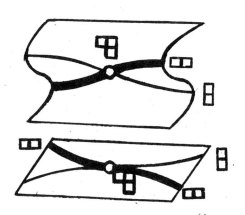

Fig. 51 - *The simplest singularities of the Schubert stratification of the Grassmannian.*

In a linear space with a fixed complete flag, we consider a special kind of one-parameter group of linear transformations; the *Jordan flow*. The simplest way to define it is to consider the model where the space is formed by polynomials in one variable, and the subspace of codimension r of the flag is formed by polynomials having at the origin a root of multiplicity at least r. In this model, the Jordan flow becomes the set of the translations of the polynomials along the axis of independent variable.

The Jordan flow acts on the Grassmann manifold. The complement to the smallest cell of this manifold is fibred into the orbits of the Jordan flow (into smooth lines). The directions of these lines may be tangent to the closure of a cell only on the boundary of the cell (formed by the adjacent cells of higher codimensions).

EXAMPLE. In fig. 51 the Jordan orbits are the vertical lines. The closure of the codimension 1 stratum is the surface $x = z^3 - 3zy^2$. The orbits are tangent to the closure along the lines $z = \pm y$, representing the closures of the two strata of codimension 2. The origin represents the stratum of codimension 3. The (x, y, z) space is a transversal submanifold of this stratum, invariant under the Jordan flow.

The projection of the trace of the train to the Jordan orbit space has a singularity along the cells of codimension larger than 1. The set of critical values is formed by two curves in the (x, y)-plane, tangent cubically at the origin to each other.

THEOREM. *In the parameter space of the families of the sequences of functions, the typical singularities of the (continuations of the) boundaries of the sets of fundamental systems coincide with the singularities of the critical value sets of the projection (to the space of Jordan orbits) of the Schubert stratification of the Grassmannian along the orbits of the Jordan flow (restricted to the submanifolds transversal to the orbits and invariant under the Jordan flow).*

EXAMPLE. The stratum of codimension 3 of the Schubert stratification, where it intersects the closures of the two strata of codimension 2, corresponds to a pair of cubically tangent plane curves (fig. 51).

The Schubert cells in the Grassmannian are numbered by Young diagrams. Let us consider the n-plane, generated by the monomials

$$t^{k_0}, t^{k_1}, \ldots, t^{k_{n-1}}, \quad \text{where} \quad k_0 < k_1 < \ldots < k_{n-1}.$$

The cell containing this point of the Grassmannian, consists of the planes generated by the polynomials with the same lower order terms. The sequence k will be called the *order of the plane* (and of the cell, and of any plane in this cell). A typical plane has *minimal order* $(0, 1, \ldots, n-1)$. The difference between the order of a plane and the minimal order is a sequence

$$(a_0 \leq a_1 \leq a_2 \leq \ldots) = (k_0 \leq k_1 - 1 \leq k_2 - 2 \leq \ldots),$$

called the *anomaly of the plane* (and of its cell). The anomalies define the *Young diagrams*, whose lines have lengths $a_{n-1} \geq \ldots \geq a_0$.

EXAMPLE. The Taylor polynomials of the functions t and $\sin t$ define a plane of order $(1, 3)$. Its *anomaly* is $(1, 2)$. The Young diagram is: ⌐⌐

The adjacency of cells is defined by the inclusion of the diagrams. The initial part of the hierarchy is shown in fig. 52.

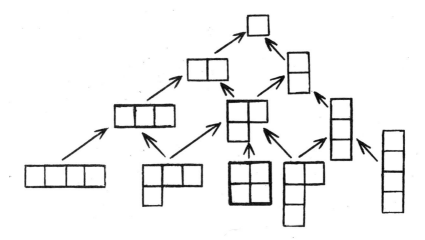

Fig. 52 - The initial part of the hierarchy of the Young diagrams.

The codimension of a cell is the "area" of the diagram, $a_0 + a_1 + \ldots$.

The zero entries in the anomaly sequence are not represented in the Young diagram. Hence the same Young diagram corresponds to planes of different dimensions n in spaces of different dimensions N. The corresponding singularities depend only on the diagram, not on n or N (if they are large enough). This holds both for the traces of the stratifield train on a k- dimensional submanifold transversal to a Schubert cell defined by a diagram of area k, and for the bifurcation diagrams (the critical sets of the projections of this trace along the Jordan orbits).

THEOREM. *The bifurcation diagrams corresponding to symmetrical Young diagrams are diffeomorphic.*

The corresponding fundamental systems consist, in general, of different numbers of functions. Hence this duality relates the differential equations of different orders to each other.

EXAMPLE. The two Young diagrams: ⊞ and ⊞ correspond to the same surface, shown in fig. 46. However the smooth part corresponds in one case to the stratum ▭, and in the other to the stratum ▯ (find which diagram corresponds to each possibility).

The pair of umbrellas shown in fig. 47, corresponds to the self-symmetrical Young diagram ⊞. The symmetry permutes the umbrellas.

Generalizing the interpretation of the pair of umbrellas provided by fig. 50, the general theory, described above, may be formulated as the theory of bifurcations of the flattenings of the projective curves (or of Weierstrass points). Let us define a curve in terms of homogeneous coordinates as the set of points $[f_1(t) : \ldots : f_n(t)]$. The Taylor polynomials of these functions at the origin define a plane in a space equipped with a complete flag. The bifurcation values of the parameters on which the projective curve depends, are the values for which the curve has an unusual flattening (more complicated than the one corresponding to the Young diagram of one point).

The *simple Young diagrams* (giving birth to a finite number of bifurcations and not to the moduli) in the Kazarijan theory are:

1) $(3, 2_k, 1_l)$;
2) $(k, 2, 1_l)$;
3) $(4, 3, 1_l)$;
4) $(3_2, 2, 1_l)$;
5) $(5, 5, 4)$;
6) the subdiagrams of these;
7) the dual diagrams to these.

Here i_j means j lines of length i.

[It seems that this list of Young diagrams has not yet appeared in other branches of Mathematics (in contrast to the list A, D, E of simple Dynkin diagrams, describing, say, the simple (= having no moduli) quivers. To a quiver there correspond the categories of linear spaces (corresponding to the points of the diagram) and linear mappings (corresponding to the oriented edges). The connected quivers, whose categories have no moduli, are the Dynkin diagrams (A, D, E) of simple (having no moduli) singularities of functions, oriented arbitrarily.]

REFERENCES

[1] V.I. ARNOLD, *Spaces of functions with critical points of bounded multiplicities*, Funkt. anal. and appl., 1989, vol. 21, **3**, pp. 1-10.
[2] V.I. ARNOLD, *Mathematical Methods of Classical Mechanics*, 2-d Ed. Springer, GTN-60, N.Y. 1989.
[3] V.A. VASSILJEV, *Lagrange and Legendre Characteristical Classes*, Gordon and Breach. N.Y. 1988.
[4] B.A. HESSIN, *Homogeneous vector fields and Whitney umbrellas*, Russian Math. Surveys, translation of the Uspehi Math. Nauk, 1987, v. 42, **5**, pp. 217-218.
[5] V.I. ARNOLD - M.B. SEVRJUK, *Oscillations and Bifurcations in Reversible Systems*, in: *Nonlinear Phenomena in Plasma Physics and Hydrodynamics*, R.Z. Sagdeev Ed. MIR 1986, pp. 31-64.
[6] M.B. SEVRJUK, *Reversible systems*, Springer Lect. Notes in Math., 1986, v. 1211, p. 320.
[7] O.P. SHCHERBAK, *Wave fronts and reflection groups*, Russian Math. Surveys, translation of the Uspehi Math. Nauk, 1988, v. 43, **3**, pp. 125-160.
[8] A.B. GIVENTAL, *Varieties of polynomials having a root of fixed comultiplicity*, Funkt. anal. and appl., 1982, v. 16, **1**, pp. 10-14.
[9] O.P. SHCHERBAK, *Projectively dual space curves and Legendre singularities*, Trudy Tbilisskogo Univ., 1982, v. 232-233, **13-14**, pp. 280-336.
[10] V.I. ARNOLD, *Lagrange varieties, asymptotical rays and the open swallowtail*, Funkt. anal. and appl., 1981, **15**, pp. 235-246.
[11] A.B. GIVENTAL, *Lagrange embeddings of surfaces and the open Whitney umbrella*, Funkt. anal. and appl., 1986, v. 20, **3**, pp. 35-41.
[12] O.A. PLATONOVA, *Projections of smooth surface*, Trudy Sem. I.G. Petrovski, 1984, vol. 10, pp. 135-149 (translated in J. Soviet Math., 1986, v. 85, **6**, pp. 2796-2808).
[13] I.G. SHCHERBAK, *Duality of boundary singularities*, Russian Math. Surveys, translation of the Uspehi Math. Nauk, 1984, v. 39, **2**, pp. 195-196.
[14] V.I. ARNOLD, *Critical points of functions on manifolds with boundary, simple Lie groups B_k, C_k, F_4 and evolutes singularities*, Russian Math. Surveys, translation of the Uspehi Math. Nauk, 1978, v. 33, **5**, pp. 91-105.

[15] I.G. SHCHERBAK, *Focal sets of surfaces with boundary and caustics of reflection groups B_k, C_k, F_4*, Funkt. anal. and appl., 1984, v. 18, 1, pp. 90-91.

[16] L.V. LEVANTOVSKII, *Singularities of the stability boundary*, Funkt. anal. and appl., 1982, v. 16, **1**, pp. 44-48.

[17] V.I. MATOV, *Ellipticity domains of generic families of homogeneous polynomials and extrema functions*, Funkt. anal. and appl., 1985, v. 19, 2, pp. 26-36.

[18] A.D. WEINSTEIN - B.Z. SHAPIRO, *Singularities of hyperbolicity domains*, Itogi Nauki i Techniki VINITI. Sovrem Probl. Math. Novejsh. Dostij. 1988, v. 33, pp. 55-78 (translated as J. Sov. Math., Consultant Bureau Plenum, N.Y.).

[19] M.F. ATIYAH - R. BOTT - L. GARDING, *Lacunas for hyperbolic differential equations with constant coefficients*, Acta Math. 1970, **124**, pp. 109-189, and 1973, **131**, pp. 145-206.

[20] V.A. VASSILJEV, *Sharpness and local Petrovskii condition for strictly hyperbolic operators with constant coefficients*, Izv. AN SSSR, ser. matem., 1986, v. 50, **2**, pp. 242-283 (translated as Sov. Math. Izvestia).

[21] V.I. ARNOLD - V.A. VASSILJEV - V.V. GORJUNOV - O.V. LJASHKO, *Singularities II. Classification and applications*, Dynamical Systems 8. Itogi nauki i techniki VINITI. Sovrem. Probl. Math. Fundamental Napravl., 1989, **39**, pp. 189-239 (§2 of ch. 4, ch. 5). (Translated as Encyclopedia of Mathematical Sciences, Springer).

[22] V.I. ARNOLD, *Hyperbolical polynomials and Vandermond mappings*, Funkt. anal. and appl., 1986, v. 20, **2**, pp. 52-53.

[23] A.N. VARCHENKO, *On normal forms of nonsmoothness of solutions of hyperbolical polynomials*. Izv. AN SSSR ser. matem., 1987, v. 51, **3**, pp. 652-665.

[24] V.I. ARNOLD, *On surfaces, defined by hyperbolical polynomials*, Math. Zametki, 1988, v. 44, **1**, pp. 3-18.

[25] A.B. GIVENTAL, *Moments of random variables and equivariant Morse Lemma*, Russian Math. Surveys, translation of the Uspehi Math. Nauk, 1987, v. 42, **2**, pp. 221-222.

[26] V.I. ARNOLD, *On the interior scattering of waves, defined by hyperbolic variational principles*, J. Geom. Phys, 1988, v. 5, **3**, pp. 305-315.

[27] V.I. ARNOLD, V.A. VASSILJEV, *Newton's Principia read 300 years later*, Notices AMS, 1989.

[28] B.Z. SHAPIRO, *Linear differential equations and manifolds of complete flags*, pp. 160-163 in the book [21] (§5 of Chapter 3).

[29] M.E. KAZARIJAN, *Singularities of the boundary of fundamental systems, flattening of projective Space curve and Schubert cells*, Itogi nauki i techniki VINITI, Sovrem. Probl. Math., Novejsh. Dostij. 1988, **33**, pp. 215-234. (Translated as J. Sov. Math., Consulant Bureau Plenum N.Y.).

[30] V.I. ARNOLD, *Singularities of the boundary of the domain of fundamental systems*, pp. 143-160 in the book [21] (§4 of Chapter 3).

INDEX

anomaly of a plane	65
bicaustic	37
bifurcation diagram	8,9
bifurcation diagram of functions	50
bifurcation diagram of zeroes	45
bifurcation hypersurface	16
bifurcation set	5
bifurcation set of zeroes	44
Bruhat ordering	58
caustic	6,7,9,13
complete flag	57
comultiplicity	34
critical point	5
critical point, non-degenerate	5
crystallographic groups	26
curvature centre set	13
degenerate field	16
disconjugate	57
discriminant	24
Dynkin diagram	26
ellipticity boundary	53
ellipticity domain	53
equivalent projection	40
evolute	13,29
focal set	13
folded umbrella	36
fundamental system	60
hyperbolicity boundary	56
hyperbolicity domain	56

icosahedron	28
invariant	23
Jordan flow	64
local hyperbolicity	56
Maxwell set	6,7
minimal order	65
Morse Lemma	5
multiplicity	56
non-crystallographic groups	26
normal form, equivalent to	45
normal mapping	13
open cell	64
order of a plane	65
perestroika	8
pinch point	16
pyralek	38
reflection group	21
regular orbit	22
Schubert cell	57,64
stability boundary	52
stability domain	52
swallowtail	9,10
Tchebyshev system	57
train	57,64
unfurled swallowtail	32
vertex	16
Vieta mapping	23
Whitney-Cayley umbrella	14
Whitney tuck mapping	11
Young diagram	65

"Pantograf" - Via alla Stazione di Voltri 2/A - Genova
Finito di stampare nel Gennaio 1991